Lecture Notes in Biomathematics

Managing Editor: S. Levin

10

John J. Tyson

The Belousov-Zhabotinskii Reaction

Springer-Verlag
Berlin · Heidelberg · New York 1976

Author

John J. Tyson
Department of Mathematics
State University of New York at Buffalo
4246 Ridge Lea Road
Amherst, N. Y. 14226/USA

Library of Congress Cataloging in Publication Data

Tyson, John J.   1947-
   The Belousov-Zhabotinskii reaction.

   (Lecture notes in biomathematics ; 10)
   Bibliography: p.
   Includes index.
   1. Belousov-Zhabotinskii reaction. 2. Differential
equations. I. Title. II. Series.
QD502.T97   541'.393   76-25897

AMS Subject Classifications (1970): 34-02

ISBN 3-540-07792-8 Springer-Verlag Berlin · Heidelberg · New York
ISBN 0-387-07792-8 Springer-Verlag New York · Heidelberg · Berlin

TABLE OF CONTENTS

PREFACE

In 1958 B.P. Belousov discovered that the oxidation of citric acid by bromate in the presence of cerium ions does not proceed to equilibrium methodically and uniformly, like most chemical reactions, but rather oscillates with clocklike precision between a yellow and colorless state. See Fig. II.1, p. 30. A.M. Zhabotinskii followed up on Belousov's original observation and in 1964 his first investigations appeared in the Russian journal Biofizika.

Though H. Degn (in Copenhagen at the time) knew of Zhabotinskii's work and published his own account of the mechanism of oscillation in Nature (1967), this interesting reaction attracted little attention among Western scientists until 1968, when Zhabotinskii and his coworkers and Busse (from Braunschweig, W. Germany) reported on their work at an international conference on biological and biochemical oscillators held in Prague.

Shortly thereafter appeared a flurry of papers on temporal oscillations and spatial patterns in this reaction system. Vavilin and Zhabotinskii (1969) [and later Kasperek and Bruice (1971)] studied the kinetics of the oxidation of Ce^{+3} by BrO_3^- and the oxidation of organic species by Ce^{+4}. Busse (1969) reported his observation of colored bands of chemical activity propagating up and down in a long tube of unstirred solution. Zaikin and Zhabotinskii (1970) observed circular chemical waves in thin layers of solution. By using an ion-specific electrode to follow the Br^- concentration, and by a thorough summary of the thermodynamics and kinetics of oxybromine species in acid solution, Field, Kőrös and Noyes (1972) were able to suggest an elaborate mechanism for temporal oscillations. A little later Bornmann, Busse and Hess (1973) published a series of illuminating articles on the overall reaction.

Meanwhile Winfree (1972) rediscovered spiral waves in thin layers of solution
(they had been previously observed by Zhabotinskii) and later (1973, 1974)
he traced the origin of several kinds of spatial patterns (spirals,
elongated rings and scroll rings) to three-dimensional scroll-shaped waves
of chemical activity. At the same time Kopell and Howard (1973) [and
independently Thoenes (1973)] suggested a purely kinematic explanation for
the band-propagation in long tubes and furthermore began a fruitful study
of wave-like solutions to reaction-diffusion equations.

As news of this pioneering work has spread, a wide variety of scientists,
mathematicians and engineers have become interested in the unusual properties
of the Bélousov-Zhabotinskii reaction. Since there are few examples of
sustained oscillations in chemical systems, especially in single phase
systems (Nicolis and Portnow, 1973), chemists are intrigued by the BZ
reaction and would like to know the mechanistic details. Unfortunately,
for the very same reason, chemists tend to dismiss the whole subject of
chemical oscillations as an academic curiosity. Biologists and biochemists,
on the other hand, are intensely interested in oscillations and pattern
formation in chemical reaction systems because of the omnipresence of clocks
and maps in living systems (Robertson and Cohen, 1972). Indeed, at the
Prague conference in 1968 it was the interest of biochemists studying glycolytic
oscillations which brought to the attention of a broad scientific audience
the early work of Zhabotinskii and others (Chance, et al., 1973). Particularly
striking similarities to the propagating chemical waves reported by Zaikin
and Zhabotinskii and by Winfree are found in the aggregation of cellular
slime molds (Robertson and Cohen, 1972), the growth pattern of fungi
(Bourret, Lincoln and Carpenter, 1969), the propagation of action potentials
(FitzHugh, 1961; Nagumo, et al., 1962), and fibrillation of heart tissue
(Gul'ko and Petrov, 1972; Krinsky, 1973). More tenuous connections can be

seen to amphibian morphogenesis, control of cell division, circadian rhythms...
(I do not mean to endorse any speculations here; I only mean to generate a
little enthusiasm!)

Mathematicians have become interested in the BZ reaction because it
provides a new field for the application of modern methods of analysis of
differential equations. The sets of many (i.e. more than two) first-order
nonlinear ordinary differential equations generated by mechanisms for single
phase, homogeneous chemical reactions present formidable challenges to the
mathematician well-versed in classical methods (Andronov, Vitt and Khaikin,
1966; Coddington and Levinson, 1955), in bifurcation theory (Marsden and
McCracken, 1975), and in topological dynamics (Hirsch and Smale, 1974).
Nonlinear parabolic partial differential equations, of the sort generated by
coupling reaction and diffusion, present more difficulties and much study is
currently being given to these problems (Kopell and Howard, 1975; Hastings,
1975). Even though there is no connection between the mechanism of the BZ
reaction and the mechanisms for the sundry biological examples mentioned
in the last paragraph, the mathematical techniques developed for the better
understood chemical problem will carry over to the biological applications,
if and when the biological mechanisms can be realistically formulated in
terms of ordinary and partial differential equations.

The Belousov-Zhabotinskii reaction has also caught the attention of
chemical engineers who, in the study of chemical reactor design, have been
interested for some time now in chemical instabilities, multiple steady state
behaviour and sustained oscillations (Schmitz, 1974).

It may be for one of these reasons that you have read thus far. In
these lecture notes, which list heavily to the mathematical side, I have
tried to present a consistent, self-contained discussion of temporal and
spatial organization in the Belousov-Zhabotinskii reaction. They are not

meant to provide a complete review of the literature to date. In Chapter I
the non-chemist will find an introduction to chemical kinetics and the non-
mathematician will find an introduction to ordinary and partial differential
equations. Chapter II outlines the mechanism of the BZ reaction in some
detail. Those not mathematically inclined may prefer to skip Chapter III,
which contains elaborate analysis of three ordinary differential equations
suggested by Field and Noyes (1974) as a simple model of the Field-Körös-
Noyes mechanism. (In the appendix may be found a similar analysis of a
different set of three ordinary differential equations suggested by
Zhabotinskii, et al. (1971) as a simple model of their mechanistic studies.)
Chapter IV presents several theories of chemical wave propagation in one,
two and three dimensions. Though this last chapter is disappointingly
incomplete, it represents the current state of the art to my knowledge.
Look for new advances here.

Let me encourage those who have not yet experimented with the Belousov-
Zhabotinskii reaction to give it a try. For your convenience I have given
recipes for producing homogeneous oscillations (p. 30) and propagating waves
(pp. 70f). The chemicals and glassware are readily available in almost any
wet-chemistry laboratory. Just ask!

A few words about notation. Probably the reader has already gathered
that original sources are referred to by giving the authors' names and date
of publication. References are listed in alphabetical order at the end of
the book. Within any given chapter, equations, figures and problems are
numbered sequentially and referred to as Eq. (1), Fig. 2, etc. Chemical
reactions are named by letters and numbers [for example, reaction (R5) or
(F)] according to the notation used originally by Field, Körös and Noyes (1972).
In referring to a figure or problem from a different chapter, I use the notation

Fig. I.1, Problem II.2, etc. To facilitate these cross references, the reader will find a Roman numeral in the upper left hand corner of each page indicating the chapter number.

Much of the material presented here was developed for a course entitled "Temporal and spatial organization in chemical systems", given at the State University of New York at Buffalo in the Spring semester, 1975. The treatment of periodic solutions of the Field-Noyes model in the relaxation oscillator regime (pp.54-69) was developed while writing up the lecture notes for publication.

I would like to thank the Research Foundation of the State University of New York for supporting me while I wrote this material. For their stimulation throughout the year, I am indebted to my colleagues at SUNY/Buffalo: Stuart Hastings, Jim Boa, Brian Hassard, Jim Greenberg and Nicholas Kazarinoff.

John J. Tyson
Amherst, New York

CHAPTER I. PRELIMINARIES

An understanding of chemical oscillations and wave patterns in the Belousov-Zhabotinskii reaction requires some familiarity with the language and methods of chemical kinetics on one hand and some facility with the mathematics of differential equations on the other. Since not every reader can be expected to know both fields to the extent which we will need later, I present in this chapter a short discussion of chemical reaction rate laws and mechanisms, and of nonlinear ordinary and partial differential equations. To strengthen the connection between this review material and the later chapters, I have drawn the examples and problems here from literature relevant to the Belousov-Zhabotinskii reaction.

Chemical kinetics

In acid solution Br^- and BrO_3^- react to form molecular bromine according to

(F) $$BrO_3^- + 5Br^- + 6H^+ = 3Br_2 + 3H_2O \quad .$$

The small integers appearing in (F) to balance atomic species are known as stoichiometric coefficients. Notice that the net charge on both sides of the reaction is balanced as well.

Chemical kinetics is the study of the rate at which such reactions proceed. The rate of reaction (F) is defined as

$$r_F = -\frac{d}{dt}[BrO_3^-] = -\frac{1}{5}\frac{d}{dt}[Br^-] = +\frac{1}{3}\frac{d}{dt}[Br_2]$$

where brackets indicate the concentration[*] of a chemical species. In order
that the rate of reaction be uniquely defined, we adopt the convention that
the rate of change of concentration of species X be divided by the
stoichiometric coefficient of species X in the balanced chemical reaction.
(We must also agree that the reaction be balanced with the smallest possible
whole-integer coefficients, and that coefficients of reactants be considered
negative integers.) Technically speaking, we have defined the rate of
change of the extent of reaction (F) .

Bray and Liebhafsky (1935) have measured the initial rate of production
of Br_2 for various initial concentrations of reactants. They found

(1)
$$r_F^{(i)} = k_F [BrO_3^-][Br^-][H^+]^2$$

where the superscript i denotes the <u>initial</u> rate. As the reaction proceeds,
the rate law becomes more complicated due to contributions of the reverse
reaction $(Br_2 \rightarrow BrO_3^- + Br^-)$.

Expression (1) is said to be a fourth order rate law, because the rate
depends on the product of four concentrations. The rate constant, k_F ,
depends on temperature and ionic strength (a measure of the overall concentration of charged species in solution). At $25°C$ in strongly acid medium,

$$k_F \approx 2\,M^{-3}\,\sec^{-1} .$$

The units of k_F are determined by the requirement that a reaction rate
always has the units $M\sec^{-1}$; see the definition of r_F .

_____

[*]Concentration can be measured in many different units. Most common is the
unit of molarity, symbolized by M . A one molar (1M) solution of chemical
X contains 6.02×10^{23} molecules of X per liter of solution.

Notice that, from the balanced chemical equation (F), one cannot deduce the rate law (1). The latter must be determined experimentally. For some simple reactions, however, the exponents in the rate law correspond exactly to the stoichiometric coefficients in the balanced equation. Such reactions are called elementary. For example, the initial rate of the gaseous reaction

$$H_2 + I_2 \rightarrow 2HI$$

is simply

$$\frac{1}{2}\frac{d}{dt}[HI] = k[H_2][I_2] \quad \text{(initially)}$$

if the system is sufficiently dilute. Under these conditions the rate of the reaction is primarily determined by the probability of a collision between a hydrogen molecule and an iodine molecule. Whereas the order of an overall reaction cannot be deduced from the balanced chemical equation, the order of an elementary reaction is just the sum of the stoichiometric coefficients of reactants. To make this distinction explicit, the order of an elementary reaction is called its molecularity.

If a reaction is not elementary, it must proceed by a series of elementary steps, known as the mechanism of the reaction. For instance, the dependence of the initial rate of overall reaction (F) on the specific combination $[BrO_3^-][Br][H^+]^2$ can be understood in terms of the mechanism

(R3) $BrO_3^- + Br^- + 2H^+ \rightarrow HBrO_2 + HOBr$

(R2) $HBrO_2 + Br^- + H^+ \rightarrow 2HOBr$

(R1) $HOBr + Br^- + H^+ \rightarrow Br_2 + H_2O$

Each of these reactions, involving the transfer of a single oxygen atom from one chemical species to another, is elementary. Their rates have been measured by various investigators (see Field, Körös and Noyes (1972) for references):

$$r_{R3}^{(i)} = k_{R3}[BrO_3^-][Br^-][H^+]^2 \ , \quad k_{R3} = 2.1\,M^{-3}\,sec^{-1}$$

$$r_{R2}^{(i)} = k_{R2}[HBrO_2][Br^-][H^+] \ , \quad k_{R2} = 2 \times 10^9\,M^{-2}\,sec^{-1}$$

$$r_{R1}^{(i)} = k_{R1}[HOBr][Br^-][H^+] \ , \quad k_{R1} = 8 \times 10^9\,M^{-2}\,sec^{-1} \ .$$

If $[BrO_3^-] = [Br^-] = [H^+] = 1M$ and $[HBrO_2] = [HOBr] = 0$ initially, then we see that reaction (R3) supplies $HBrO_2$ and $HOBr$ at the rate $2M\,sec^{-1}$. Within a fraction of a second, $[HBrO_2]$ builds up to $10^{-9}M$ and the rate of reaction (R2) is also $2M\,sec^{-1}$. Were $[HBrO_2]$ to increase further, (R2) would proceed faster than (R3) and $[HBrO_2]$ would decrease back to $10^{-9}M$. Were $[HBrO_2]$ to drop below $10^{-9}M$, the opposite would occur. Thus reactions (R2) and (R3) quickly establish a "pseudo-steady state" concentration of $HBrO_2$,

$$[HBrO_2] = \frac{k_{R3}}{k_{R2}}[BrO_3^-][H^+] \ ,$$

such that the flux through both reactions is identical. Similarly, reactions (R1) and (R3) establish

$$[HOBr] = \frac{k_{R3}}{k_{R1}}[BrO_3^-][H^+] \ .$$

Thus we see that step (R3) is the bottleneck in process (F) = (R3) + (R2) + 3(R1) and controls the overall flux:

$$r_F^{(i)} = -\frac{d}{dt}[BrO_3^-] = r_{R3}^{(i)} = 2M^{-3}\,sec^{-1}[BrO_3^-][Br^-][H^+]^2 \ .$$

It is important to recognize that reaction mechanisms are educated guesses. The chemist hypothesizes a mechanism to explain a measured rate law, then he tests further predictions of the mechanism. As experimental evidence accumulates, the field of possible mechanisms narrows. However, even if a mechanism is falsified, it often remains as a useful model under certain conditions. A classic example is the $H_2 + I_2$ reaction mentioned earlier (see Sullivan, 1967).

The principles of thermodynamics assure us that all chemical reactions are reversible. For example,

(-R1)
$$Br_2 + H_2O \rightarrow HOBr + Br^- + H^+$$

$$r_{-R1}^{(i)} = k_{-R1}[Br_2] \quad , \quad k_{-R1} = 10^2 \, sec^{-1} \quad .$$

In dilute aqueous solution, $[H_2O] = 55.5\,M$ always, and this constant has been absorbed into k_{-R1} by convention. For any given concentrations of Br_2, $HOBr$, Br^-, and H^+, the rate of (R1) is

$$r_{R1} = \frac{d}{dt}[Br_2] = r_{R1}^{(i)} - r_{-R1}^{(i)}$$

$$= k_{R1}[HOBr][Br^-][H^+] - k_{-R1}[Br_2]$$

Eventually reaction (R1) reaches equilibrium

$$r_{R1}^{(i)} = r_{-R1}^{(i)} \quad , \quad r_{R1} = 0$$

or

$$\frac{[Br_2]}{[HOBr][Br^-][H^+]} = \frac{k_{R1}}{k_{-R1}} = 8 \times 10^7 \, M^{-2} = K_{R1} \quad .$$

K_{R1} is called the equilibrium constant of reaction (R1). It varies with temperature, pressure, ionic strength, etc., along with the rate constants.

By definition, the equilibrium constant of reaction

$$aA + bB \ldots = pP + qQ + \ldots$$

is

$$K = \frac{[P]^p [Q]^q \ldots}{[A]^a [B]^b \ldots} \; .$$

For an elementary reaction, the equilibrium constant is the ratio of the forward rate constant to the reverse rate constant. For overall reaction (F) to be at equilibrium, each elementary step must be at equilibrium:

$$k_{R3} [BrO_3^-][Br^-][H^+]^2 = k_{-R3}[HBrO_2][HOBr]$$

$$k_{R2} [HBrO_2][Br^-][H^+] = k_{-R2}[HOBr]^2$$

$$k_{R1} [HOBr][Br^-][H^+] = k_{-R1}[Br_2]$$

$$K_F = \frac{[Br_2]^3}{[BrO_3^-][Br^-]^5[H^+]^6} = \frac{k_{R3}}{k_{-R3}} \frac{k_{R2}}{k_{-R2}} \left(\frac{k_{R1}}{k_{-R1}} \right)^3$$

With the further information that $k_{-R3} = 10^4 M^{-1} sec^{-1}$, $k_{-R2} = 5 \times 10^{-5} M^{-1} sec^{-1}$ we calculate

$$K_F = 10^{34} M^{-9} .$$

Thus,

$$\frac{[Br_2]^3}{[BrO_3^-][Br^-]^5} = 10^{34}[H^+]^6 M^{-9} = \begin{cases} 10^{28} M^{-3} & \text{at} \quad pH = 1 \\ 10^{-2} M^{-3} & \text{at} \quad pH = 6 \\ 10^{-26} M^{-3} & \text{at} \quad pH = 10 \end{cases}$$

At equilibrium, molecular bromine is in great abundance over the ionic species, bromate and bromide, in acidic solution. In neutral or basic solution, this distribution is reversed.

To illustrate the experimental determination of rate constants and mechanisms, we turn to another set of reactions important in the mechanism of the Belousov-Zhabotinskii reaction. The overall reaction for the bromination of malonic acid is

(R8) $$Br_2 + CH_2(COOH)_2 = BrCH(COOH)_2 + Br^- + H^+ \ .$$

This occurs by a two step mechanism

(R8a) [structure: malonic acid, H₂C(COOH)₂] $\xrightarrow{\text{enolization}}$ [enol structure] (reversible)

(R8b) [enol structure] $+ Br_2 \xrightarrow{\text{bromination}}$ [brominated structure] $+ Br^- + H^+$

If the rate of bromination is much faster than the rate of enolization, then the rate of (R8) will be limited by the first step

(2) $$r_{R8}^{(i)} = -\frac{d}{dt}[MA] = k_a[MA] \ , \quad MA = CH_2(COOH)_2 \ .$$

If bromination is the slow step, then (R8a) will equilibrate

$$k_a[MA] = k_{-a}[enol] \Rightarrow [enol] = K_a[MA] \ , \quad K_a = \frac{k_a}{k_{-a}} \ ,$$

and (R8b) will limit the rate of the overall reaction

(3) $$r_{R8}^{(i)} = k_b[enol][Br_2] = k_b K_a[MA][Br_2] \ .$$

Thus we can distinguish between these two possibilities by determining whether (R8) is governed by a first order or second order rate law.

Problem 1. Let $\alpha = [MA]_0$, $\beta = [Br_2]_0$, the initial concentrations of malonic acid and molecular bromine. Let $x(t) = [BrMA]$, the concentration of bromomalonic acid at time t . Then

$$[MA] = \alpha - x \ , \quad [Br_2] = \beta - x \ .$$

Show that, if (R8a) is rate limiting,

$$\ln \frac{\alpha - x}{\alpha} = -k_a t \ ;$$

and that, if (R8b) is rate limiting,

$$\ln \left(\frac{\alpha}{\beta} \cdot \frac{\beta - x}{\alpha - x} \right) = (\beta - \alpha) k_b K_a t \ .$$

From the following experimental data of West (1924), show that the bromination of malonic acid is rate-limited by the enolization step. Evaluate k_a .

(Ans. $k_a = .0085 \ min^{-1}$)

t (min)	[MA] (M)	[Br$_2$] (M)
0	0.0300	0.00661
1.92	0.0289	0.00650
3.83	0.0278	0.00639
5.92	0.0267	0.00628

The bromination of malonic acid does not stop with bromomalonic acid but goes on to dibromomalonic acid:

(R8′) $Br_2 + BrCH(COOH)_2 = Br_2C(COOH)_2 + Br^- + H^+$.

Again this is a two step process: first enolization, then bromination.

Problem 2. From West's data below, show that the production of dibromomalonic acid is rate-limited by the bromination step. Evaluate $k_b' K_a'$. The equilibrium constant can be determined thermodynamically, and thus the rate constant is specified. (Ans. $k_b' K_a' = 4.17 M^{-1} min^{-1}$)

t (min)	[BrMA] = [Br$_2$] (M)
0	0.01446
2.5	0.01253
4.2	0.01157
6.05	0.01060
8.29	0.00964
11.2	0.00868
14.3	0.00771
19.1	0.00674

For further details on chemical kinetics, consult any textbook on physical chemistry, e.g. Daniels and Alberty (1966).

Ordinary differential equations

We have seen that the rate of a reaction in a homogeneous (well-stirred) solution at constant temperature and pressure* is naturally expressed as a differential equation for the time rate of change of chemical concentrations. In Problem 1 the reader was called upon to integrate simple first order and second order rate laws. In this section we will develop techniques for dealing with more complicated differential equations: linear and nonlinear, with two or more dependent variables.

When several reactions proceed simultaneously, we generate systems of

---

*As long as gaseous reactions are not under consideration, volume changes upon reaction are very small and justly neglected.

first order ordinary differential equations. Consider, for example, the sequential isomerization reactions

(I)
$$A \underset{k_{-1}}{\overset{k_1}{\rightleftharpoons}} B \underset{k_{-2}}{\overset{k_2}{\rightleftharpoons}} C \quad .$$

Let $a = [A]$, $b = [B]$, $c = [C]$, $\cdot = d/dt$. For definiteness, let $k_1 = 4 \sec^{-1}$, $k_{-1} = 3.5 \sec^{-1}$, $k_2 = 4.5 \sec^{-1}$, $k_{-2} = 2 \sec^{-1}$. Then

$$\dot{a} = -4a + 3.5b$$
$$\dot{b} = 4a - 8b + 2c$$
$$\dot{c} = 4.5b - 2c \quad .$$

These equations are not all independent since $\dot{a} = \dot{b} = \dot{c} = 0$, that is,

(4)
$$a + b + c = m \ , \ \text{a constant.}$$

This is just a statement of conservation of mass. Our reaction system can be described completely by Eqs. (4) and (5).

(5)
$$\dot{a} = -4a + 3.5b$$
$$\dot{b} = 2a - 10b + 2m$$

At equilibrium, the forward and reverse rates of each reaction are identical. In this case, this is exactly equivalent to the statement that $\dot{a} = \dot{b} = 0$. We have for equilibrium concentrations

$$a^* = 7m/33 \ , \quad b^* = 8m/33 \quad .$$

Starting from arbitrary initial conditions, satisfying $a(0) + b(0) + c(0) =$ does the system eventually come to equilibrium? Let us show that the deviations from equilibrium

$$x = a - a^* \ , \quad y = b - b^*$$

approach 0 as $t \to \infty$. In terms of x , y , Eq. (5) is

(6)
$$\dot{x} = -4x + 3.5y$$
$$\dot{y} = 2x - 10y$$

or in matrix form

(7)
$$\dot{x} = Kx$$

where in this case $x = (x,y)$ is a 2×1 vector and K is a 2×2 matrix. In general we can interpret x as an $n \times 1$ vector and K as an $n \times n$ matrix.

Systems of linear ordinary differential equations (7) can always be solved in terms of the eigenvectors and eigenvalues of the matrix K .[#] For, if $\boldsymbol{\xi}$ is an eigenvector of K , with eigenvalue λ , that is,

$$K\boldsymbol{\xi} = \lambda\boldsymbol{\xi} , \quad \lambda = \text{a scalar} ,$$

then direct substitution verifies that

(8)
$$x = e^{\lambda t}\boldsymbol{\xi}$$

is a solution of system (7). If K has n linearly independent eigenvectors (for instance, if all the eigenvalues are distinct), then the general solution of the system of DE (7) can be expressed as an arbitrary linear combination of fundamental solutions (8); that is

$$x(t) = \sum_{i=1}^{N} c_i e^{\lambda_i t} \boldsymbol{\xi}_i .$$

The constants c_i are determined by the initial condition, $x(0)$.

[#]See almost any undergraduate textbook on differential equations, e.g. Boyce and Di Prima (1969, chap. 7).

Problem 3. Solve system (6) subject to the arbitrary initial conditions $x(0) = \alpha$, $y(0) = \beta$.

Ans.

$$\begin{pmatrix} x \\ y \end{pmatrix} = \frac{2\alpha + \beta}{16} e^{-3t} \begin{pmatrix} 7 \\ 2 \end{pmatrix} + \frac{2\alpha - 7\beta}{16} e^{-11t} \begin{pmatrix} 1 \\ -2 \end{pmatrix} .$$

From the solution given to Problem 3, we see that $(x,y) \rightarrow (0,0)$ as $t \rightarrow \infty$ for arbitrary initial conditions. That is, the system of isomers, A , B , C , will indeed equilibrate regardless of initial conditions.

Problem 4. Consider $\dot{x} = Kx$, where

$$x = \begin{pmatrix} x_1 \\ x_2 \\ x_3 \end{pmatrix} , \quad K = \begin{pmatrix} 10 & 10 & 0 \\ 0 & -1 & 14.4 \\ 1 & 0 & -1 \end{pmatrix} .$$

Show that as $t \rightarrow +\infty$

$$x(t) \rightarrow e^{11t} \begin{pmatrix} 1.44 \\ 0.144 \\ 0.12 \end{pmatrix} .$$

Problem 5. Solve

(a) $\dot{x} = \begin{pmatrix} -1 & -1 \\ \frac{1}{2} & 0 \end{pmatrix} x$ and (b) $\dot{x} = \begin{pmatrix} -1 & -1 \\ \frac{1}{2} & 0 \end{pmatrix} x + \begin{pmatrix} 1 \\ 0 \end{pmatrix} .$

Ans.

(a) $x_h(t) = c_1 e^{-\tau} \begin{pmatrix} 2\cos\tau \\ \sin\tau - \cos\tau \end{pmatrix} + c_2 e^{-\tau} \begin{pmatrix} -2\sin\tau \\ \sin\tau + \cos\tau \end{pmatrix}$, $\tau = t/2$.

(b) $x(t) = x_h(t) + x_p(t)$, $x_p(t) = \begin{pmatrix} 0 \\ 1 \end{pmatrix}$.

In general, chemical reaction systems cannot be described by linear differential equations, as in (7). Rather, rate laws appear as products of dependent variables (chemical concentrations), most commonly quadratic terms generated by bimolecular steps in the reaction mechanism. For the case of a well-stirred solution, we are concerned with systems of nonlinear ordinary differential equations

$$(9) \qquad\qquad \dot{x} = f(x)$$

where $x \in R^n$ and $f : R^n \to R^n$ is a nonlinear vector-valued function. If $x = (x_1, \ldots, x_n)$ represent chemical concentrations, then we are only interested in solutions of (9) which lie in the positive orthant $R^{+n} = \{(x_1, \ldots, x_n) \mid x_i \geq 0 , 1 \leq i \leq n\}$. If x_i represents a deviation from equilibrium, then $x_i < 0$ is permitted.

Most often the simplest solutions of (9) to be found are constant, or steady state, solutions

$$x(t) = x_0 , \text{ a constant vector such that}$$

$$f(x_0) = 0 .$$

For chemical systems which exchange matter with an external bath, there may be several steady state solutions of Eq. (9).

Problem 6. Consider the mechanism (Edelstein, 1970)

(E1) $$A + X \rightleftharpoons 2X$$

(E2) $$X + E \rightleftharpoons Y$$

(E3) $$Y \rightleftharpoons E + B .$$

(E1) describes the autocatalytic production of X from A, and steps (E2), (E3) describe the enzymatic degradation of X to B. (E = enzyme, Y = enzyme-substrate complex, $[E] + [Y]$ = total enzyme concentration = a constant.) Assume that A and B are maintained at constant concentrations, $[A] = a$, $[B] = b$, by exchange with an external bath. (Chemically, this can be achieved in some instances by buffering.) For simplicity let all rate constants $= 1$. Show that reaction system (E) can be described by two differential equations

$$\dot{x} = ax - x^2 - x(e_T - y) + y$$

(10)

$$\dot{y} = x(e_T - y) - 2y + b(e_T - y)$$

where $x = [X]$, $y = [Y]$, $e = [E]$, $e_T = [E] + [Y] = $ constant . Under what conditions will there be three steady state solutions of (10) in the positive quadrant: $x > 0$, $y > 0$.

Once we have found the chemically significant steady state solutions of (9), it is informative to determine the stability of these solutions. Do small perturbations away from the steady state damp out or grow larger with time?

Let $y = x - x_0$, and expand $f(x)$ in a Taylor series about x_0 . Eq. (9) becomes

$$\dot{y} = f(x_0 + y) = f(x_0) + f_x(x_0) \cdot y + y \cdot f_{xx}(x_0) \cdot y + \ldots$$

Using the fact that $f(x_0) \equiv 0$ and keeping only the dominant terms for $|y| \ll 1$, we are left with a system of linear differential equations

(11) $$\dot{y} = Ly , \quad L = f_x(x_0)$$

which describe approximately the behavior of solutions of (9) in a neighborhood of the constant solution, x_0 . The matrix, L , appearing in (11) is just the Jacobian of $f(x)$ evaluated at $x = x_0$,

$$L_{ij} = \frac{\partial f_i}{\partial x_j} (x_1^0, \ldots, x_n^0) \ .$$

Now we have just mentioned that the solutions of (11) can be expressed in terms of the eigenvalues and eigenvectors of L. If all the eigenvalues of L have real part less than zero, then $y = x - x_0 \to 0$ as $t \to +\infty$, i.e. the steady state is asymptotically stable since arbitrary perturbations die out as $t \to +\infty$. However, if any eigenvalue of L has positive real part, then there exist perturbations which increase as t increases, i.e. the steady state is unstable.

Problem 7. (a) In Edelstein's example, introduced in the last problem, with $a = 8.49$, $b = 0.2$, $e_T = 30$, there are three steady states: $(1.0, 11.25)$, $(1.7, 14.6)$, $(3.6, 19.65)$. Show that the first and third are asymptotically stable but the second is unstable.

Ans. For the first steady state,

$$L = \begin{pmatrix} -12.26 & 2.0 \\ 18.75 & -3.2 \end{pmatrix}$$

with eigenvalues -15.33 and -0.13. Since both eigenvalues are real and negative, the steady state is asymptotically stable.

(b) Can you extend the local picture generated in part (a) to a global picture (which is "qualitatively" correct) of solutions of (10) in the positive quadrant? Hints:

(i)   trajectories, $\{(x(t), y(t)) \,|\, -\infty < t < +\infty\}$, never cross;

(ii)   for structurally stable ("coarse") dynamical systems in the plane, R^2, as $t \to \pm\infty$, a trajectory must either diverge to infinity, or approach a constant solution, or approach a periodic solution,

$\{(x(t),y(t)) \mid x(t+T) = x(t), \; y(t+T) = y(t) \;$ for some $\; T > 0 \;,$

and $\; (x(t),y(t)) \neq$ constant solution$\}$;

(iii) for Eq. (10) it can be shown that there are no periodic solutions. Property (i) is a trivial consequence of the existence-uniqueness theorem for autonomous ordinary differential equations. A nice discussion of coarse systems in the plane (ii) can be found in Andronov, Vitt and Khaikin (1966, see pp. 396-398 in particular). Property (iii) is established by application of Bendixson's negative criterion (Andronov, Vitt and Khaikin, 1966, p. 305). An example of the "qualitative" solution of nonlinear ordinary differential equations in the plane, similar to the one studied in this problem, is worked out in the textbook of Boyce and DiPrima (1969, pp. 398-401).

To illustrate further the analysis of nonlinear differential equations, consider the mechanism due to Prigogine and Lefever (1968)

(P1) $\qquad\qquad\qquad\qquad\qquad A \rightarrow X$

(P2) $\qquad\qquad\qquad\qquad B + X \rightarrow Y + D$

(P3) $\qquad\qquad\qquad\qquad 2X + Y \rightarrow 3X$

(P4) $\qquad\qquad\qquad\qquad\qquad X \rightarrow E \qquad .$

We assume that the reactions are all irreversible and that the concentrations of species A and B are maintained constant. No confusion should result from using the same symbol for a chemical species and its concentration. The rate equations for intermediates, X and Y , are

$$\frac{dX}{dt} = k_1 A - k_2 BX + k_3 X^2 Y - k_4 X$$

$$\frac{dY}{dt} = k_2 BX - k_3 X^2 Y .$$

These can be considerably simplified by introducing the dimensionless variables

$$x = \sqrt{\frac{k_3}{k_4}} X , \quad y = \sqrt{\frac{k_3}{k_4}} Y , \quad \tau = k_4 t$$

and parameters

$$a = \frac{k_1 A}{k_4} \sqrt{\frac{k_3}{k_4}} , \quad b = \frac{k_2 B}{k_4} .$$

In which case,

(12)

$$\frac{dx}{d\tau} = a - bx + x^2 y - x$$

$$\frac{dy}{d\tau} = bx - x^2 y .$$

There is one and only one constant solution of (12), $x_0 = a$, $y_0 = b/a$. What is its stability? Let $\xi = x - x_0$, $\eta = y - y_0$, and linearize (12):

$$\frac{d}{d\tau} \begin{pmatrix} \xi \\ \eta \end{pmatrix} = \begin{pmatrix} b - 1 & a^2 \\ -b & -a^2 \end{pmatrix} \begin{pmatrix} \xi \\ \eta \end{pmatrix} .$$

The eigenvalues of the Jacobian matrix are

$$\lambda = \frac{(b - 1 - a^2) + \sqrt{(b - 1 - a^2)^2 - 4a^2}}{2} .$$

The steady state (x_0, y_0) is stable if and only if

$$b < 1 + a^2 .$$

What happens when the steady state is unstable, when $b > 1 + a^2$? When (x_0, y_0) is unstable, no trajectories can approach this solution as $t \to +\infty$. Since this is the only constant solution of (12), we see that (recalling hint (ii) of problem 7) trajectories must go either to infinity or to a periodic solution as $t \to +\infty$. If we can show that solutions of (12) are bounded, then we can be sure that there exists at least one periodic solution of (12). (Tyson, 1973).

To show that solutions of (12) are bounded, we will show that any half-trajectory, $\{(x(t), y(t)) | 0 \le t \le +\infty\}$, which starts in a certain domain $\mathcal{D} \subset R^2$ must remain in $\mathcal{D}$. More precisely, if $(x(0), y(0)) \in \overline{\mathcal{D}}$, then $(x(t), y(t)) \in \mathcal{D}$ for all $t > 0$, where $\overline{\mathcal{D}}$ is the closure of the open set $\mathcal{D}$. We go through the argument in some detail because similar reasoning will be used later to prove the existence of periodic solutions for a model of the Belousov-Zhabotinskii reaction.

Consider the set

$$\mathcal{D} = \{(x, y) | x > 0 , y > 0 , y < \alpha - x , y < \beta + x\}$$

where $\alpha > 0$, $\beta > 0$ are as yet unspecified. See Fig. 1. $\overline{\mathcal{D}}$ is defined similarly with $\ge$, $\le$ signs. The boundary of $\mathcal{D}$, consists of sections of the lines: $x = 0$, $y = 0$, $y = \alpha - x$, $y = \beta + x$.

Let $\hat{\imath} = (1, 0)$, the unit vector along x axis, $\hat{\jmath} = (0, 1)$, the unit vector along y axis, and

$$\vec{v} = \hat{\imath} \frac{dx}{dt} + \hat{\jmath} \frac{dy}{dt} = \left(\frac{dx}{dt} , \frac{dy}{dt} \right) ,$$

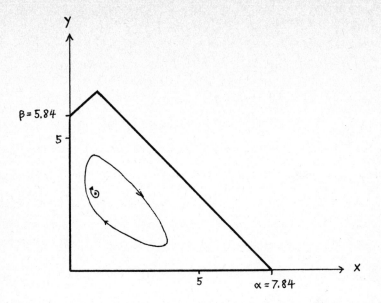

Fig. 1.   System (12) for   a = 1 ,   b = 3 .   The unstable steady state, the
stable limit cycle (calculated numerically), and the boundary of
Ω   are illustrated.

the vector field defined by (12).   We want to show that the vector field
$\vec{v}$ is directed into   Ω   at every point on   ∂Ω , that is,

$$\vec{v} \cdot \vec{n} < 0 \quad \text{on} \quad \partial\Omega$$

where the vector $\vec{n}$ is normal to   ∂Ω   and directed outward.

On that section of the boundary for which   x = 0 , $\vec{n}$ = (-1,0)   and

$$\vec{v} \cdot \vec{n} \Big|_{x=0} = - \frac{dx}{d\tau}\Big|_{x=0} = -a < 0 \quad .$$

When   y = 0 , $\vec{n}$ = (0,-1)   and

$$\vec{v} \cdot \vec{n}\Big|_{y=0} = - \frac{dy}{d\tau}\Big|_{y=0} = -bx < 0 \quad \text{for} \quad x > 0 \ .$$

At $x = 0$, $y = 0$, the vector field $\vec{v}$ is tangent to $\partial\mathcal{D}$; however, since $dx/d\tau = a > 0$ and $d^2y/d\tau^2 = ab > 0$ at the origin, no trajectory can leave $\overline{\mathcal{D}}$ through the origin. For, if $(x(0), y(0)) = (0,0)$, then $x(t) > 0$, $y(t) > 0$ over some interval $0 < t < \epsilon$.

When $y = \alpha - x$, $\vec{n} = (1,1)$ and

$$\vec{v} \cdot \vec{n}\Big|_{y=\alpha-x} = \frac{dx}{d\tau} + \frac{dy}{d\tau}\Big|_{y=\alpha-x} = a - x < 0 \quad \text{for} \quad x > a \ .$$

Thus for any $\alpha > 0$, $y = \alpha - x$ is a satisfactory boundary as long as $x > a$ We propose to use $y = \beta + x$ as the boundary for $x < a$. Then $\vec{n} = (-1,1)$,

$$\vec{v} \cdot \vec{n}\Big|_{y=\beta+x} = - \frac{dx}{d\tau} + \frac{dy}{d\tau}\Big|_{y=\beta+x} = -a + 2bx - 2x^2y + x < 0 \quad ?$$

In order that $-a + (2b+1)x - 2x^2y < 0$, we need

$$(13) \qquad\qquad\qquad y > \frac{(2b+1)x - a}{2x^2} \ .$$

The right hand side of (13) has a maximum value $(2b+1)^2/8a$ achieved at $x = 2a/(2b+1)$. For $b > 1 + a^2$, $x = 2a/(2b+1) < a$, so we are in the correct region. We must choose β so that the line $y = \beta + x$ goes through the point $x = 2a/(2b+1)$, $y = (2b+1)^2/8a$. Thus

$$\beta = \frac{(2b+1)^3 - 16a^2}{8a(2b+1)} \ .$$

Finally, we choose α such that when $x = a$, the two equations, $y = \alpha - a$ and $y = \beta + a$, give the same value of y . Thus

$$\alpha = 2a + \beta \ .$$

Problem 8.   Prove that $\beta > 0$, if $b > 1 + a^2$.

This completes the proof that trajectories of (12) are bounded (for $b > 1 + a^2$), which implies the existence of at least one periodic solution of (12) whenever the steady state is unstable.   See Fig. 1.   Periodic solutions for this particular model have been studied extensively (Nicolis, 1971; Lavenda, et. al., 1971; Tyson, 1973).

For a complete, concise treatment of nonlinear oscillations, see Minorsky (1964).

Reaction-diffusion equations.

If a reaction mixture is not well stirred, diffusion tends to maintain spatial homogeneity; but, since diffusion is a slow process, spatial in-homogeneity may persist for long periods of time.   When coupled with certain chemical reactions, diffusion may also be responsible for maintaining stable spatial patterns.

For systems with a single thermodynamically stable phase, chemical species flow in a direction to decrease any concentration gradients.   For dilute solutions and shallow concentration gradients, the flux of chemical U is simply proportional to the gradient of its concentration u :

$$\vec{j} = -D \, \vec{\nabla} \, u \quad ,$$

where $[\vec{j}]$ = molecules $cm^{-2}\ sec^{-1}$ and $[D]$ = $cm^2 sec^{-1}$.

The diffusion constant, D , is related to the viscosity of the medium,

η , and the size of the diffusing particle, r = radius , by a formula due to Einstein

$$D = \frac{kT}{6\pi\eta r} \quad .$$

The viscosity of water at $300\,^{\circ}K$ is $\eta \approx 10^{-2} \text{g cm}^{-1} \text{sec}^{-1}$, and the radius of small molecules is $r \approx 2 \times 10^{-8} \text{cm}$, so

$$D \approx \frac{1.4 \times 10^{-16} \text{erg deg}^{-1} \times 300 \text{ deg}}{6 \times 3.14 \times 10^{-2} \text{g cm}^{-1} \text{sec}^{-1} \times 2 \times 10^{-8} \text{cm}} \approx 10^{-5} \text{cm}^2 \text{sec}^{-1} \quad .$$

This will be a handy number to remember.

If several chemicals, $U_1, U_2, \ldots, U_n$, are diffusing simultaneously, one has

$$\vec{\underset{\sim}{j}} = -\underline{\underline{D}} \, \vec{\nabla} \, \underset{\sim}{u} \quad ,$$

where $\vec{\underset{\sim}{j}}$ and $\underset{\sim}{u}$ are n-vectors and $\underline{D}$ is an $n \times n$ matrix. Off-diagonal elements of D are strongly dependent on concentration, $D_{ij} \sim u_i/u_j$, whereas diagonal elements D_{ii} are only weakly dependent on u_i . Throughout this manuscript we shall assume that $D_{ij} = D_i \delta_{ij}$, δ_{ij} = Kronecker delta, D_i = constant.

To minimize clumsy notation, we will no longer distinguish scalars u from n-vectors $\underset{\sim}{u}$, but simply write u for the vector of chemical concentrations in an n-component system. It may be that $n = 1$, but in most cases we will be dealing with multicomponent systems.

In a closed system without chemical reaction, conservation of mass implies that

(14) $$\frac{\partial u}{\partial t} = D \nabla^2 u$$

where ∇^2 is the Laplacian operator in one, two or three dimensions. With chemical reactions or sources and sinks of material, we have

(15)
$$\frac{\partial u}{\partial t} = D\nabla^2 u + f(u) \quad .$$

Problem 9.   Show that

$$u(\vec{r},t) = u(0)\exp(-r^2/4Dt) \cdot (4\pi Dt)^{-3/2}$$

satisfies the diffusion equation $u_t = D\nabla^2 u$ and the initial condition $u(\vec{r},0) = u(0)\delta(\vec{r})$, $\delta(\vec{r}) = $ Dirac delta function.   Show that

$$u(\vec{r},t) = u(0)\exp(kt)\exp(-r^2/4Dt) \cdot (4\pi Dt)^{-3/2}$$

satisfies the reaction-diffusion equation $u_t = D\nabla^2 u + ku$ with the same delta function initial condition.

Problem 10.   Solve $u_t = Du_{xx}$ subject to the boundary conditions

$$\frac{\partial u}{\partial x} = 0 \quad \text{at} \quad x = 0 , \quad x = 1$$

and the initial condition $u(x,0) = 2u_0(1 - x)$.

Ans.

$$\frac{u(x,t)}{u_0} = 1 + \sum_{n=1,3,5,\ldots} \left(\frac{2}{n\pi}\right)^2 \cos(n\pi x)\exp\{-(n\pi)^2 Dt\} \quad .$$

Nonlinear reaction-diffusion equations (15) are difficult to solve.   For instance, even the simple "epidemic" equation (Fisher, 1937)

$$\frac{\partial u}{\partial t} = D \frac{\partial^2 u}{\partial x^2} + u(1-u) \ , \qquad u = u(x,t)$$

has a long and distinguished history (Kolmogorov, Petrovsky and Piscounoff, 1937; Aronson and Weinberger, 1975). In this section we will be content to deal with reaction-diffusion equations linearized around a homogeneous steady state solution. That is, suppose the reaction equations alone, $du/dt = f(u)$, have a steady state solution, $u(\vec{r},t) = u_0$. As in the last section, let $v = u - u_0$ and expand $f(u) = f(u_0 + v) = f(u_0) + f_u(u_0) \cdot v + \dots$. Let

$$L = f_u(u_0) = \left(\frac{\partial f_i}{\partial u_j} (u_0) \right)$$

the Jacobian of $f(u)$ evaluated at $u = u_0$. On neglecting higher order terms, Eq. (15) becomes

(16)
$$\frac{\partial v}{\partial t} = D \nabla^2 v + Lv \ .$$

We solve (16) by the method of separation of variables. First, determine the eigenfunctions of the Laplacian operator:

(17)
$$\nabla^2 X_k(\vec{r}) = -k^2 X_k(\vec{r}) \ , \qquad k \in \mathcal{K} \ .$$

The eigenvalue spectrum, $\mathcal{K}$, is determined by boundary conditions. Then write $v(\vec{r},t)$ as

(18)
$$v(\vec{r},t) = X_k(\vec{r}) T(t)$$

where T is an N-dimensional vector function of time, whereas $X_k(\vec{r})$ is a scalar function of space. Substituting (18) into (16) and using (17), we obtain

$$X_k \frac{dT(t)}{dt} = -k^2 X_k DT + X_k LT \ ,$$

or, on dividing through by $X_k(\vec{r})$,

(19)
$$\frac{dT}{dt} = (L - k^2 D)T \quad .$$

Now Eq. (19) is just a set of linear, first-order ordinary differential equations, which we learned in the last section to solve in terms of the eigenvalues and eigenvectors of the $n \times n$ matrix $L - k^2 D$,

$$(L - k^2 D)\xi_{ik} = \lambda_{ik}\xi_{ik} \; , \qquad i = 1,\ldots,n \; ; \qquad k \in \mathcal{K} \quad .$$

If for each k the eigenvalues λ_{ik} are distinct, then solutions of (16) can be expressed in general as a linear combination

$$v(\vec{r},t) = \sum_{k \in \mathcal{K}} \sum_{i=1}^{n} c_{ik} X_k(\vec{r}) e^{\lambda_{ik} t} \xi_{ik} \; ,$$

where the expansion coefficients, c_{ik} , are determined by the initial conditions. If $\mathrm{Re}\,\lambda_{ik} > 0$ for any (i,k) , then arbitrary perturbations from the homogeneous steady state will grow as time increases. That is, the homogeneous steady state solution is unstable with respect to perturbations of spatial frequency k .

Problem 11. Consider a one dimensional reaction vessel $(\vec{r} = x)$ closed at the ends, $x = 0$, $x = L$. That is,

$$\frac{\partial u}{\partial x} = 0 \quad \text{at} \quad x = 0 \; , \; x = L \quad \text{for all} \quad t \quad .$$

What are the appropriate eigenfunctions of the Laplacian $\nabla^2 = d^2/dx^2$?
Ans.

$$X_k(x) = \cos kx \; , \qquad k = \frac{n\pi}{L} \; , \qquad n = 0,1,2,\ldots \quad .$$

A very instructive example has been presented by DeSimone, Beil and Scriven (1973). They consider two chemicals, U_1 and U_2, reacting in an infinite two-dimensional (i.e. thin layer of) solution:

$$\frac{\partial u}{\partial t} = D \nabla^2 u + f(u) \ , \qquad u = \begin{pmatrix} u_1 \\ u_2 \end{pmatrix} , \qquad \nabla^2 = \frac{\partial^2}{\partial r^2} + \frac{1}{r} \frac{\partial}{\partial r} + \frac{1}{r^2} \frac{\partial^2}{\partial \theta^2} \ .$$

Since they are looking for solutions of radial symmetry (target patterns, or spirals), they work in polar coordinates. Furthermore, they assume that

$$D = \begin{pmatrix} 0 & 0 \\ 0 & D \end{pmatrix} ,$$

that is, U_1 is immobile (or, in other words, U_2 diffuses much more rapidly than U_1).

Problem 12. Find the eigenfunctions of the Laplacian in polar coordinates,

$$\left(\frac{\partial^2}{\partial r^2} + \frac{1}{r} \frac{\partial}{\partial r} + \frac{1}{r^2} \frac{\partial^2}{\partial \theta^2} \right) X(r,\theta) = -k^2 X(r,\theta) \ .$$

Hint: use the method of separation of variables, i.e. write $X(r,\theta) = R(r) \, \Theta(\theta)$
Then

$$R(r) = J_m(kr) \ , \qquad Y_m(kr) \ , \quad \text{Bessel functions, and}$$

$$\Theta(\theta) = \cos m\theta \ , \ \sin m\theta \ , \qquad m = 0,1,2,\ldots \ .$$

Since we have not imposed boundary conditions on $R(r)$, k is unrestricted.

Let

$$L = f_u(u_0) = \begin{pmatrix} a & b \\ c & d \end{pmatrix} \ .$$

Then (19) becomes

(19′)
$$\frac{dT}{dt} = \begin{pmatrix} a & b \\ c & d - k^2 D \end{pmatrix} T \ .$$

The eigenvalues of $L - k^2 D$ are roots of

$$\lambda^2 - (a + d - k^2 D)\lambda + ad - bc - ak^2 D = 0 \ .$$

We want solutions which neither grow nor decay with time, so we insist that

(20)
$$k^2 D = a + d > 0 \ ,$$

in which case $\lambda = \pm i\omega$, where

(21)
$$\omega^2 = -a^2 - bc > 0 \ .$$

From Eq. (20) we see that at least one of the chemicals must be produced autocatalytically at the steady state, i.e.

$$\frac{\partial}{\partial u_i} \left(\frac{du_i}{dt} \right)_{u=u_0} > 0 \quad \text{for} \quad i = 1 \text{ or } 2 \ .$$

From Eq. (21) we see that cross-coupling terms, b and c, must be of opposite sign.

Under conditions (20) and (21) we find two fundamental solutions of (19′):

$$T_1(\omega t) = \begin{pmatrix} A \cos(\pm \omega t - \psi) \\ \cos(\pm \omega t) \end{pmatrix} \ , \quad T_2(\omega t) = \begin{pmatrix} A \sin(\pm \omega t - \psi) \\ \sin(\pm \omega t) \end{pmatrix}$$

where $A = \sqrt{-b/c} > 0$, $\psi = \tan^{-1}(\mp \omega/a)$. Combining this with the results of Problem 12, we can form the linear combination $(v = u - u_0)$

$$v(r, \theta, t) = J_m(kr) \cos m\theta \, T_1(\omega t) \pm J_m(kr) \sin m\theta \, T_2(\omega t)$$

$$- Y_m(kr) \cos m\theta \, T_1(\omega t) \mp Y_m(kr) \sin m\theta \, T_2(\omega t)$$

or

(22)
$$v(r,\theta,t) = M_m(kr) \left(\begin{array}{c} A \cos[\alpha_m(kr) \pm (\omega t \pm m\theta) - \psi] \\[2ex] \cos[\alpha_m(kr) \pm (\omega t \pm m\theta)] \end{array} \right)$$

where

$$M_m(kr) = [J_m^2(kr) + Y_m^2(kr)]^{\frac{1}{2}}$$

$$\alpha_m(kr) = \tan^{-1}[Y_m(kr)/J_m(kr)] \quad .$$

Eq. (22) becomes invalid chemically as $r \to 0$, since $M_m(kr) \to \infty$ as $r \to 0$.

Using the asymptotic expansions of J_m and Y_m

$$\left\{ \begin{array}{c} J_m \\ Y_m \end{array} \right\}(kr) \sim \sqrt{\frac{2}{\pi kr}} \left\{ \begin{array}{c} \cos \\ \sin \end{array} \right\} \left(kr - \frac{m\pi}{2} - \frac{\pi}{4} \right) , \quad \text{as} \quad r \to \infty ,$$

we find that $(r_0 \equiv (2m+1)\pi/4k)$

$$v_2 \sim \frac{2}{\sqrt{\pi kr}} \cos[k(r - r_0) \pm (\omega t \pm m\theta)] , \quad \text{as} \quad r \to \infty .$$

The amplitude of the disturbance from the steady state solution dies out like $r^{-\frac{1}{2}}$ as $r \to \infty$, and the curves of constant phase (i.e. of constant concentrati approach

(23)
$$k(r - r_0) \pm (\omega t \pm m\theta) = \text{constant} , \quad \text{as} \quad r \to \infty \quad .$$

For $m = 0$, we see from Eq. (23) that, as $r \to \infty$, the curves of constant concentration, $u(r,\theta,t) = \text{constant}$, are large circles propagating outward or inward at speed ω/k, and wavelength $2\pi/k$, where k and ω are defined in Eqs. (20) and (21) in terms of the diffusion constant and chemical kinetic parameters.

For $m = 1$, a curve of constant concentration appears asymptotically as an Archimedean spiral of pitch $1/k$ wound on a circle of radius $r_0 = 3\pi/4k$. As t increases, the whole spiral pattern rotates at angular velocity ω. Depending on the ambiguous signs in (23), the spiral is counterclockwise contracting, clockwise contracting, counterclockwise expanding, or clockwise expanding. The radial velocity is $\overline{+}\omega/k$ as $r \to \infty$. For $m > 1$, the constant phase curves are a set of m spirals of pitch m/k, asymptotically, rotating at angular velocity ω/m.

Target patterns $(m = 0)$ and single spiral waves $(m = 1)$ have been observed in thin layers of "Z" (for Zhabotinskii) reagent. We will come back to a discussion of these patterns in Chapter IV.

CHAPTER II. CHEMISTRY OF THE BELOUSOV-ZHABOTINSKII REACTION.

Oscillations in the Belousov-Zhabotinskii reaction are easily produced. A convenient recipe is given by Field (1972). Ingredients:

		initial concentrations
150 ml	1 M H_2SO_4	1 M
0.175 g	$Ce(NO_3)_6(NH_4)_2$	0.002 M
4.292 g	$CH_2(COOH)_2$	0.28 M
1.415 g	$NaBrO_3$	0.063 M

In a beaker equipped with stirring apparatus, dissolve malonic acid and cerium ammonium nitrate in sulphuric acid. Solution will first be yellow, then, after a few minutes, turn clear. When clear, add sodium bromate. Solution will turn yellow, then clear...then yellow, then clear..., oscillating with a period on the order of one minute, depending on the rate of stirring among other things. A more dramatic color change, between red and blue, can be produced by adding a few mls of 0.025 M Ferroin (1,10 phenanthroline iron). All of these chemicals are readily available.

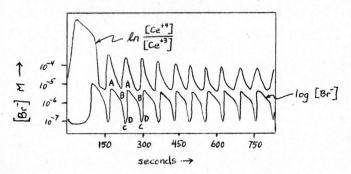

Fig. 1. Potentiometric traces (from Field, Körös and Noyes, 1972). Initial conditions: $[CH_2(COOH)_2]_0 = 0.13$ M , $[KBrO_3]_0 = 0.063$ M , $[Ce(NH_4)_2(NO_3)_5]_0 = 0.005$ M , $[H_2SO_4]_0 = 0.8$ M .

Potentiometric traces of $\log[Br^-]$ and $\log[Ce^{+4}]/[Ce^{+3}]$ are given in Fig. 1. These recordings were made by Field, Körös and Noyes (1972) under slightly different initial conditions. Since reagent grade $NaBrO_3$ is contaminated by $NaBr$, there is a little Br^- present initially. BrO_3^- and Br^- quickly come into equilibrium with $HBrO_2$, $HOBr$ and Br_2 via reactions (R1) - (R3), as discussed in Chapter I. After about 100 sec, $[Br^-]$ increases and $[Ce^{+4}]$ decreases dramatically. (Large amounts of Ce^{+4} give the solution a pale yellow color. With Ferroin as indicator, blue corresponds to the oxidized state, Ce^{+4} and Fe^{+3}, and red to the reduced state, Ce^{+3} and Fe^{+2}.) There are four distinct phases of the oscillation: AB = slow consumption of Br^-, BC = rapid consumption of Br^-, CD = slow regeneration of Br^-, DA = rapid regeneration of Br^-.

Overall reaction.

The major reactants are bromate (BrO_3^-) and malonic acid $(CH_2(COOH)_2 \equiv MA)$. They are used up slowly and monotonically. In 1 M sulphuric acid, hydrogen ions (H^+) are present in great excess and buffered by the bisulphate ion (HSO_4^-), so $[H^+]$ is not changed appreciably by the reaction. Cerium is a catalyst, that is, it facilitates the conversion of reactants to products without being itself transformed.

As just mentioned, in acid solution BrO_3^- soon comes into equilibrium with $HBrO_2$, $HOBr$, Br^-, Br_2 by a series of oxygen atom transfers (Field, Körös and Noyes, 1972).

(R3) $BrO_3^- + Br^- + 2H^+ = HBrO_2 + HOBr$

(R2) $HBrO_2 + Br^- + H^+ = 2HOBr$

(R1) $HOBr + Br^- + H^+ = Br_2 + H_2O$.

Br_2 cannot be detected in solution (Zhabotinskii, 1964), because it reacts quickly with malonic acid to form bromomalonic acid (BrMA) and dibromomalonic acid (Br_2MA). (See Chapter I, pp. 6-8.)

(R8) $\qquad Br_2 + CH_2(COOH)_2 \rightarrow BrCH(COOH)_2 + Br^- + H^+$

(R8') $\qquad Br_2 + BrCH(COOH)_2 \rightarrow Br_2C(COOH)_2 + Br^- + H^+$.

Dibromomalonic acid is unstable in acid medium, decarboxylating to dibromoacetic acid (Br_2Ac)

$$Br_2C(COOH)_2 \rightarrow Br_2CHCOOH + CO_2$$

which is further brominated to tribromoacetic acid (Br_3Ac)

$$Br_2 + Br_2CHCOOH \rightarrow Br_3CCOOH + Br^- + H^+ .$$

In this manner a number of different brominated organic compounds are produced. Ce^{+4} oxidizes these species, along with malonic acid, producing Br^- and CO_2 (Vavilin and Zhabotinskii, 1969; Field, Körös and Noyes, 1972):

(R9) $\qquad 6Ce^{+4} + CH_2(COOH)_2 + 2H_2O \rightarrow 6Ce^{+3} + HCOOH + 2CO_2 + 6H^+$

(R10) $\qquad 4Ce^{+4} + BrCH(COOH)_2 + 2H_2O \rightarrow 4Ce^{+3} + Br^- + HCOOH + 2CO_2 + 5H^+$

etc.

No formic acid (HCOOH) is detected in the reaction mixture (Bornmann, Busse and Hess, 1973b). It may be oxidized by Br_2 (Smith, 1972) producing more Br^- and CO_2 , e.g.

(R11) $\qquad Br_2 + HCOOH \rightarrow 2Br^- + CO_2 + 2H^+$.

The regeneration of Br^- is very important for the oscillation. Lots of CO_2 is produced during the reaction, as witness all the bubbles.

Unless there is so little malonic acid compared to bromate that all organic matter is finally oxidized to carbon dioxide, the final resting place for reduced bromine is in brominated organic species. By thin-layer chromatography, Bornmann, Busse and Hess (1973a) have determined BrMA and Br_2Ac as the major products. They (..., 1973b) write the overall reactions as

(0) $\qquad 2BrO_3^- + 3CH_2(COOH)_2 + 2H^+ \rightarrow 2BrCH(COOH)_2 + 3CO_2 + 4H_2O$

(0′) $\qquad 2BrO_3^- + 2CH_2(COOH)_2 + 2H^+ \rightarrow Br_2CHCOOH \quad + 4CO_2 + 4H_2O$.

They also estimate that

$$\frac{\text{extent of reaction (0)}}{\text{extent of reaction (0′)}} = \frac{3}{1} \quad .$$

Finally let us notice that BrO_3^- can be reduced by Ce^{+3} through a series of free radical intermediates (Field, Körös and Noyes, 1972)

(R5) $\qquad\qquad BrO_3^- + HBrO_2 + H^+ \rightarrow 2BrO_2 + H_2O$

(R6) $\qquad\qquad Ce^{+3} + BrO_2 + H^+ \rightarrow Ce^{+4} + HBrO_2$.

Adding (R5) and $2 \times$ (R6) we get

(G) $\qquad\qquad 2Ce^{+3} + BrO_3^- + HBrO_2 + 3H^+ \rightarrow 2Ce^{+4} + 2HBrO_2 + H_2O$.

Problem 1. Step (R5) is not elementary but involves the unstable intermediate Br_2O_4

(R5a) $\qquad\qquad BrO_3^- + HBrO_2 + H^+ = Br_2O_4 + H_2O$

(R5b) $\qquad\qquad\qquad Br_2O_4 = 2BrO_2$.

From the rate constants

$$k_{R5a} = 10^4 \, M^{-2} \, sec^{-1} \, , \qquad k_{-R5a} = 10^6 \, sec^{-1}$$

$$k_{R5b} = 10^8 \, sec^{-1} \, , \qquad k_{-R5b} = 10^9 \, M^{-1} \, sec^{-1}$$

justify the neglect of species Br_2O_4 .

Table 1. The FKN mechanism (Field, Körös and Noyes, 1972)

(R1)　　$HOBr + Br^- + H^+ \rightleftharpoons Br_2 + H_2O$

$k_{R1} = 8 \times 10^9 \, M^{-2} \, sec^{-1}$

$k_{-R1} = 110 \, sec^{-1}$

(R2)　　$HBrO_2 + Br^- + H^+ \rightarrow 2HOBr$

$k_{R2} = 2 \times 10^9 \, M^{-2} \, sec^{-1}$

(R3)　　$BrO_3^- + Br^- + 2H^+ \rightarrow HBrO_2 + HOBr$

$k_{R3} = 2.1 \, M^{-3} \, sec^{-1}$

(R4)　　$2HBrO_2 \rightarrow BrO_3^- + HOBr + H^+$

$k_{R4} = 4 \times 10^7 \, M^{-1} \, sec^{-1}$

(R5)　　$BrO_3^- + HBrO_2 + H^+ \rightleftharpoons 2BrO_2 + H_2O$

$k_{R5} = 1.0 \times 10^4 \, M^{-2} \, sec^-$

$k_{-R5} = 2 \times 10^7 \, M^{-1} \, sec^{-1}$

(R6)　　$BrO_2 + Ce^{+3} + H^+ \rightleftharpoons HBrO_2 + Ce^{+4}$

(fast)

(R8)　　$Br_2 + CH_2(COOH)_2 \rightarrow BrCH(COOH)_2 + Br^- + H^+$

$$r_{R8} = 1.3 \times 10^{-2} \, M^{-1} \, sec^{-1} [H^+][MA]$$

(R9)　　$6Ce^{+4} + CH_2(COOH)_2 + 2H_2O \rightarrow 6Ce^{+3} + HCOOH + 2CO_2 + 6H^+$

$$r_{R9} = \frac{8.8 \times 10^{-2} \, sec^{-1} [Ce^{+4}][MA]}{0.53 \, M + [MA]}$$

(R10)　$4Ce^{+4} + BrCH(COOH)_2 + 2H_2O \rightarrow 4Ce^{+3} + Br^- + HCOOH + 2CO_2 + 5H^+$

$$r_{R10} = \frac{1.7 \times 10^{-2} \, sec^{-1} [Ce^{+4}][BrMA]}{0.20 \, M + [BrMA]}$$

(R11)　$Br_2 + HCOOH \rightarrow 2Br^- + CO_2 + 2H^+$　$r_{R11} = 7.5 \times 10^{-3} \, sec^{-1} [Br_2][HCOOH]/[$

The FKN mechanism.

The reactions which we have discussed so far are collected in Table 1 along with rate data, as known.

Field, Körös and Noyes (1972) explain the oscillation as follows. When $[Br^-]$ is large, BrO_3^- is reduced by Br^- to Br_2 :

$$(R3) + (R2) + 3(R1) = (F) \qquad BrO_3^- + 5Br^- + 6H^+ \rightarrow 3Br_2 + 3H_2O \ .$$

The Br_2 reacts immediately with MA according to (R8) . Altogether

$$(F) + 3(R8) = (A) \qquad BrO_3^- + 2Br^- + 3CH_2(COOH)_2 + 3H^+ \rightarrow 3BrCH(COOH)_2 + 3H_2O \ .$$

As discussed in Chapter I (p. 4), the rate of (F) is limited by the first step (R3) . Since (R8) is fast with respect to (R3) , the rate of process (A) is

$$(1) \qquad r_A = r_F = r_{R3} = -\frac{d[BrO_3^-]}{dt} = k_{R3}[BrO_3^-][Br^-][H^+]^2 \ .$$

When $[Br^-]$ is small, BrO_3^- is reduced by Ce^{+3}

$$(R5) + 2(R6) = (G) \qquad 2Ce^{+3} + BrO_3^- + HBrO_2 + 3H^+ \rightarrow 2Ce^{+4} + 2HBrO_2 + H_2O \ .$$

The rate limiting step for process (G) is (R5) , that is

$$r_G = -\frac{d[BrO_3^-]}{dt} = +\frac{d[HBrO_2]}{dt} = k_{R5}[BrO_3^-][HBrO_2][H^+] \ .$$

Bromous acid is produced autocatalytically; its concentration grows exponentially. Eventually this growth is limited by the disproportionation reaction

$$(R4) \qquad 2HBrO_2 \rightarrow HOBr + BrO_3^- + H^+ \ .$$

The net effect of $2(G) + (R4)$ is process

(B) $\qquad BrO_3^- + 4Ce^{+3} + 5H^+ \rightarrow HOBr + 4Ce^{+4} + 2H_2O$.

As long as $[Br_2]$ is small, process (B) is rate limited by the first step, (R5) :

(2) $\qquad r_B = \frac{1}{2} r_G = \frac{1}{2} r_{R5} = \frac{1}{2} k_{R5}[BrO_3^-][H^+][HBrO_2]$,

since, for every BrO_3^- lost in process (B) , two must be used up in reaction (G) .

Problem 2. When process (B) is dominant, the rate equation for bromous acid is

$$\frac{d[HBrO_2]}{dt} = k_{R5}[BrO_3^-][H^+][HBrO_2] - 2k_{R4}[HBrO_2]^2 \ .$$

By scaling $[HBrO_2]$ and t , put this equation in dimensionless form

$$\frac{du}{d\tau} = u(1 - u) \qquad \text{- the "logistic" equation}$$

and solve for $u = u(\tau)$. Convert your answer back to $[HBrO_2](t)$.
Ans.

$$[HBrO_2]_t = \frac{k_{R5}[BrO_3^-][H^+][HBrO_2]_0}{2k_{R4}[HBrO_2]_0 - (2k_{R4}[HBrO_2]_0 - k_{R5}[BrO_3^-][H^+])\exp(-k_{R5}[BrO_3^-][H^+]t)}$$

It is rewarding to consider processes (A) and (B) as alternate fates for $HBrO_2$. If $[Br^-]$ is large, $[HBrO_2]$ is kept small by step (R2) and Br^- is slowly consumed by process (A) . (This corresponds to the "slow bromide consumption" phase, AB , in Fig. 1.) When $[Br^-]$ drops to a critical concentration (point B in Fig. 1), the autocatalytic production of $HBrO_2$ from BrO_3^- , step (R5), overtakes the consumption of $HBrO_2$ by Br^- , step (R2). At this point, $[HBrO_2]$ grows explosively, $[Br^-]$ decreases abruptly because of reaction (R2) , and much of the Ce^{+3} is converted to Ce^{+4} . (Compare phase BC in Fig. 1.)

The critical Br^- concentration is determined by the competition between steps (R2) and (R5) :

$$\frac{d[HBrO_2]}{dt} \sim -k_{R2}[HBrO_2][Br^-][H^+] + k_{R5}[BrO_3^-][HBrO_2][H^+]$$

$$\sim \{k_{R5}[BrO_3^-] - k_{R2}[Br^-]\}[HBrO_2]$$

$$\sim -[HBrO_2] \quad \text{if} \quad k_{R5}[BrO_3^-] < k_{R2}[Br^-]$$

$$\sim +[HBrO_2] \quad \text{if} \quad k_{R5}[BrO_3^-] > k_{R2}[Br^-] \quad .$$

Thus

(3) $$[Br^-]_{crit} = \frac{k_{R5}}{k_{R2}} [BrO_3^-] = 5 \times 10^{-6} [BrO_3^-] \quad .$$

In the experiment recorded in Fig. 1, $[BrO_3^-] = 0.063\,M$, so

$$[Br^-]_{crit} = 0.3 \times 10^{-6}\,M$$

which agrees well with the measured $[Br^-]$ at point B .

Br^- is regenerated by the oxidation of brominated organic compounds by Ce^{+4} , as discussed earlier (p. 31). For instance, adding (R10) + (R11) + (R1) , we get process

(C) $HOBr + 4Ce^{+4} + BrCH(COOH)_2 + H_2O \rightarrow 2Br^- + 4Ce^{+3} + 3CO_2 + 6H^+$.

Ce^{+4} is converted back to Ce^{+3} and $[Br^-]$ increases (phase CD in Fig. 1). The critical Br^- concentration at which the system switches back from process (B) to process (A) (at point D in Fig. 1) is smaller than $[Br^-]_{crit}$ in Eq. (3), because $[Ce^{+4}]/[Ce^{+3}]$ is much larger at point D than it is at point B . Field, Körös and Noyes discuss this on p. 8661 of their paper.

The oxidation of organic species by Ce^{+3} to release Br^- is undoubtedly more complex than process (C) . There are other brominated organic species in solution (Br_2Ac , Br_3Ac) . We do not know the stoichiometry involved, that is, the number of Br^- ions released per Ce^{+4} consumed. Nor do we have much detail about the kinetics of this part of the reaction (Vavilin and Zhabotinskii, 1969; Kasperek and Bruice, 1971). So, for the meantime, we will use reaction (C) as a convenient formulation of the feedback process, recognizing that some revision may be necessary later.

Notice that the sum of processes (A), (B) and (C) is the overall (C) reaction (0) .

$$2(A) + 3(B) + (C) = 3(0) \ .$$

Problem 3. Account for overall reaction $(0')$ in terms of the chemistry discussed in this section.

In review: During phase AB (see Fig. 1), $[Br^-] \sim 5 \times 10^{-6} > [Br^-]_{crit}$. Process (A) is proceeding at a rate given by Eq. (1), with $[BrO_3^-] = .063\,M$, $[H^+] = 1.05\,M$

$$r_A = k_{R3}[BrO_3^-][Br^-][H^+] \approx 10^{-6} - 10^{-7}\,M\,sec^{-1} \ , \text{ during phase AB} \ .$$

Br^- is consumed at twice this rate by process (A), but at the same time Br^- is being produced by process (C). There is a net loss of Br^- during phase AB.

When process (A) dominates, $HBrO_2$ is primarily produced by (R3) and destroyed by (R2), so that

$$[HBrO_2] \sim \frac{k_{R3}}{k_{R2}} [BrO_3^-][H^+] \sim 6 \times 10^{-11} M, \text{ during phase AB .}$$

From this value, we compute from Eq. (2)

$$r_B \sim 2 \times 10^{-8} M \sec^{-1}, \text{ during phase AB .}$$

At point B the system switches from process (A) to process (B). $[HBrO_2]$ jumps to a very large value (see Problem 2)

$$[HBrO_2] = \frac{k_{R5}[BrO_3^-][H^+]}{2k_{R4}} \sim 8 \times 10^{-6} M .$$

$[Br^-]$ decreases abruptly because of reaction (R2). The potentiometer reading is $10^{-7} M$, but $[Br^-]$ may be considerably lower. From Eqs. (1) and (2)

$$r_A < 10^{-8} M \sec^{-1}, \quad r_B \sim 2 \times 10^{-3} M \sec^{-1}, \text{ during phase CD .}$$

Process (B) produces lots of Ce^{+4}, which oxidizes the brominated organic species, releasing Br^-. Thus, during phase CD, $[Br^-]$ increases.

At point D, $[Br^-]$ is large enough so that step (R2) overtakes (R5) as the principal fate of $HBrO_2$. The system switches back to process (A). $[HBrO_2]$ drops precipitously because of step (R4), and $[Br^-]$ quickly increases because the Br^- sink, step (R2), shuts off as $[HBrO_2]$ drops, whereas the Br^- source, process (C), continues producing. At point A the source and sink of Br^- balance each other, and then $[Br^-]$ begins to decrease along with $[Ce^{+4}]$.

This sounds reasonable enough, but one might equally expect that the competing processes (A), (B), (C) reach a kinetic steady state, in mutual balance. This indeed happens under conditions for which oscillations are not observed. However, by constructing an elaborate computer model simulating twenty reactions, Edelson, Field and Noyes (1975) found chemical oscillations of the type just described.

In the next chapter we will investigate the properties of a much simpler version of the FKN mechanism, one simple enough to study analytically.

CHAPTER III. THE OREGONATOR.

From our discussion in the last chapter, we can isolate five important steps from the   FKN   mechanism:

(R3) $$BrO_3^- + Br^- + 2H^+ \rightarrow HBrO_2 + HOBr$$

(R2) $$HBrO_2 + Br^- + H^+ \rightarrow 2HOBr$$

(G) $$2Ce^{+3} + BrO_3^- + HBrO_2 + 3H^+ \rightarrow 2Ce^{+4} + 2HBrO_2 + H_2O$$

(R4) $$2HBrO_2 \rightarrow BrO_3^- + HOBr + H^+$$

(R10) $$4Ce^{+4} + BrCH(COOH)_2 + 2H_2O \rightarrow 4Ce^{+3} + Br^- + HCOOH + 2CO_2 + 5H^+ \quad .$$

Step   (R3)   is rate-limiting for process   (A) .   Step   (R2)   is important in switching control from process   (A)   to process   (B) .   Reaction   (G)   represents the autocatalytic production of $HBrO_2$ in process   (B) .   It is not elementary: the rates of reaction   (G)   and process   (B)   are limited by step   (R5) .

(R5) $$BrO_3^- + HBrO_2 + H^+ \rightarrow 2BrO_2 + H_2O \quad .$$

Step   (R4)   limits the growth of $HBrO_2$.   Step   (R10)   initiates the regeneration of Br^- from brominated organic species, which we have called process   (C) .

The model, steady states and stability

Field and Noyes (1974a) invented a kinetic model, which they called the "Oregonator", based on the five steps just mentioned:

(M1) $A + Y \rightarrow X + P$

(M2) $X + Y \rightarrow 2P$

(M3) $A + X \rightarrow 2X + 2Z$

(M4) $2X \rightarrow A + P$

(M5) $Z \rightarrow hY$.

With $A = BrO_3^-$, $P = HOBr$, $X = HBrO_2$, $Y = Br^-$ and $Z = Ce^{+4}$, steps (M1) , (M2) and (M4) are simply (R3) , (R2) and (R4) , step (M3) has the stoichiometry of (G) and the kinetics of (R5) , and step (M5) represents the regeneration of Br^- at the expense of Ce^{+4} (h is an unspecified stoichiometric coefficient). As I have written scheme (M) , it differs slightly from the Oregonator as originally proposed by Field and Noyes. In their scheme, $Z = 2[Ce^{+4}]$ and their stoichiometric parameter f is twice the number of Br^- formed per Ce^{+4} used up in the feedback process.

Assume that reactions (M) proceed in a well-stirred solution at constant temperature and pressure. Furthermore, assume that the depletion of A (BrO_3^- can be neglected over times on the order of minutes (Noyes, 1976). Finally assume that the reactions are all irreversible, in which case the product, P , has no affect on the kinetics (Field, 1975). Then the time rate of change of the intermediates, X , Y and Z , is given by (see Chapter I , pp. 1 - 4)

$$\frac{dX}{dt} = k_{M1}AY - k_{M2}XY + k_{M3}AX - 2k_{M4}X^2$$

$$\frac{dY}{dt} = -k_{M1}AY - k_{M2}XY + hk_{M5}Z$$

$$\frac{dZ}{dt} = 2k_{M3}AX - k_{M5}Z \quad .$$

From Table 1 we have, since $[H^+] \sim 1M$ in $0.8M\ H_2SO_4$,

$$k_{M1} = k_{R3}[H^+]^2 \sim 2\,M^{-1}\,sec^{-1}$$

$$k_{M2} = k_{R2}[H^+] \sim 2 \times 10^9\,M^{-1}\,sec^{-1}$$

$$k_{M3} = k_{R5}[H^+] \sim 10^4\,M^{-1}\,sec^{-1}$$

$$k_{M4} = k_{R4} \sim 4 \times 10^7\,M^{-1}\,sec^{-1}$$

$$k_{M5} = 4r_{R10}[Ce^{+4}]^{-1} \sim 0.4[BrMA]\,M^{-1}\,sec^{-1}, \text{ for } [BrMA] \ll 0.2M$$

$h \sim 0.5$ from the stoichiometry of process (C), but see Noyes and Jwo (1975) for a fuller discussion.

The rate equations are more manageable in dimensionless form. Let

$$\xi = (k_{M2}/k_{M1}A)X \sim (2 \times 10^{10}\,M^{-1})\,[HBrO_2]$$

$$\eta = (k_{M2}/k_{M3}A)Y \sim (3 \times 10^6\,M^{-1})\,[Br^-]$$

$$\rho = (k_{M2}k_{M5}/2k_{M1}k_{M3}A^2)Z \sim (6 \times 10^3\,M^{-1})\,[Ce^{+4}]$$

$$\tau = (k_{M1}A)t \sim (0.1\,sec^{-1})t$$

$$\epsilon = k_{M1}/k_{M3} = k_{R3}[H^+]/k_{R5} \sim 2 \times 10^{-4}$$

$$p = k_{M1}A/k_{M5} = k_{R3}[H^+]^2[BrO_3^-]/0.4[BrMA] \sim 300$$

$$q = 2k_{M1}k_{M4}/k_{M2}k_{M3} = 2k_{R3}k_{R4}/k_{R2}k_{R5} \sim 8 \times 10^{-6} \quad .$$

e have used $[H^+] = 1\,M$, $[BrO_3^-] = 0.06\,M$, $[BrMA] = 10^{-3}\,M$ (see Edelson, t. al., 1975) in the numerical estimates. In terms of these variables and arameters, the rate equations for mechanism (M) are

$$\epsilon \frac{d\xi}{d\tau} = \xi + \eta - q\xi^2 - \xi\eta$$

1)
$$\frac{d\eta}{d\tau} = -\eta + 2h\rho - \xi\eta$$

$$p\frac{d\rho}{d\tau} = \xi - \rho \quad .$$

What can we say about solutions of Eq. (1)?

There is a trivial solution $\xi = \eta = \rho = 0$.

There are two other steady state solutions defined by

(2a)
$$\rho = \xi$$

(2b)
$$\eta = 2h\xi(1 + \xi)^{-1}$$

(2c) $\quad q\xi^2 + (\eta - 1)\xi - \eta = 0$.

Combining (2b) and (2c) ,

(3)
$$\xi = \frac{1 - 2h - q \pm \sqrt{(1 - 2h - q)^2 + 4q(2h + 1)}}{2q}$$

Only the positive root has chemical significance. By (ξ_0, η_0, ρ_0) we will denote this steady state in the positive octant.

Problem 1. Derive the following approximations for (ξ_0, η_0, ρ_0) as $q \to 0$:

(4a) $\quad h < \frac{1}{2}$: $\qquad \rho_0 = \xi_0 \approx (1 - 2h)/q \qquad , \quad \eta_0 \approx 2h$

(4b) $\quad h \approx \frac{1}{2}$: $\qquad \rho_0 = \xi_0 \approx \sqrt{2/q} \qquad , \quad \eta_0 \approx h + \frac{1}{2}$

(4c) $\quad h > \frac{1}{2}$: $\qquad \rho_0 = \xi_0 \approx (2h + 1)/(2h - 1) , \quad \eta_0 \approx h + \frac{1}{2}$.

Is (ξ_0, η_0, ρ_0) stable or unstable with respect to small perturbations? As in Chapter I, we let

$$x = \xi - \xi_0 , \qquad y = \eta - \eta_0 , \qquad z = \rho - \rho_0 .$$

Then, using a dot to denote differentiation with respect to τ ,

$$\epsilon \dot{x} = -\alpha x - \beta y - qx^2 - xy$$

(5)
$$\dot{y} = -\gamma x - \delta y + 2hz - xy$$

$$p\dot{z} = x - z$$

where

$$\alpha = -1 + \eta_0 + 2q\xi_0 = q\xi_0 + \frac{\eta_0}{\xi_0} > 0$$

$$\beta = \xi_0 - 1 > 0 \ , \quad \text{for} \quad q < 1$$

(6)
$$\gamma = \eta_0 > 0$$

$$\delta = \xi_0 + 1 > 0 \ .$$

If $|x| \ll 1$, $|y| \ll 1$, $|z| \ll 1$, we can neglect the quadratic terms on the right hand side of Eq. (5) and only consider the linearized equations

(5L)
$$\dot{x} = Kx \ , \qquad x = \begin{pmatrix} x \\ y \\ z \end{pmatrix} \ , \qquad K = \begin{pmatrix} -\alpha/\epsilon & -\beta/\epsilon & 0 \\ -\gamma & -\delta & 2h \\ 1/p & 0 & -1/p \end{pmatrix} \ .$$

As discussed in Chapter I, (5L) can be solved in terms of the eigenvalues and eigenvectors of the matrix K . The characteristic equation of K is

(7)
$$0 = - \begin{vmatrix} -\alpha/\epsilon - \lambda & -\beta/\epsilon & 0 \\ -\gamma & -\delta - \lambda & 2h \\ 1/p & 0 & -1/p - \lambda \end{vmatrix} = \lambda^3 + \alpha\lambda^2 + \beta\lambda + \mathcal{C}$$

where

$$\mathcal{a} = \frac{\alpha}{\epsilon} + \delta + \frac{1}{p} > 0$$

$$\mathcal{B} = \frac{\alpha\delta}{\epsilon} + \frac{\delta}{p} + \frac{\alpha}{p\epsilon} - \frac{\beta\gamma}{\epsilon}$$

$$C = \frac{\alpha \delta}{p \epsilon} + (2h - \gamma) \frac{\beta}{p \epsilon} > 0 \ , \ \text{since} \ 2h - \gamma = 2h (1 + \xi_0)^{-1} \ .$$

Since $\alpha > 0$ and $C > 0$, at least one eigenvalue is real and negative.[*] β can be of either sign. If $\beta \gg 0$, then the eigenvalues are

$$\lambda_1 \approx -C/\beta \ , \quad \lambda_{2,3} \approx -\alpha/2 \pm i \sqrt{\beta} \ .$$

Since $\text{Re} \ \lambda_i < 0$, $1 \leq i \leq 3$, small perturbations from (ξ_0, η_0, ρ_0) die out as $\tau \to +\infty$, i.e. the steady state is stable. If $\beta \ll 0$, then

$$\lambda_1 \approx C/|\beta| \ , \quad \lambda_{2,3} \approx \pm \sqrt{|\beta|} \ .$$

Since there are two real positive eigenvalues, an arbitrary perturbation from (ξ_0, η_0, ρ_0) will grow exponentially with increasing time, i.e. the steady state is unstable. If $\beta = C/\alpha$, then

$$\lambda_1 \approx -\alpha \ , \quad \lambda_{2,3} = \pm i \sqrt{\beta} \ .$$

---

[*]This follows from Descartes' rule of signs (Burnside and Panton, 1928), which is an extremely handy tool for doing linear stability analysis. Consider the polynomial equation

(#) $$x^n + a_1 x^{n-1} + a_2 x^{n-1} + \ldots + a_{n-1} x + a_n = 0 \ .$$

Let N = the number of sign changes in the sequence $(1, a_1, a_2, \ldots, a_n)$, ignoring any zeroes. Then there are at most N real positive roots of (#). Furthermore, there are exactly either N, or $N-2$, or $N-4$, or $\ldots$ real positive roots. For example, consider

($) $$x^3 + ax^2 - bx + c = 0 \ \text{where} \ a > 0 \ , \ b > 0 \ , \ c > 0 \ .$$

There are either 2 or 0 real positive roots of ($). Let $y = -x$. Then ($) becomes

(%) $$y^3 - ay^2 - by - c = 0 \ .$$

There is exactly one real positive root of (%), and thus exactly one real negative root of ($).

Re $\lambda_{2,3} = 0$; i.e. the steady state is marginally stable. A more rigorous
analysis (Murray, 1974a),using the Routh-Hurwitz criteria (Gantmacher, 1959),
shows that $\alpha\beta = C$ is indeed the boundary between stable and unstable
behavior close to (ξ_0, η_0, ρ_0) . For $\epsilon = 10^{-4}$, $q = 10^{-5}$ the relation
$\alpha\beta = C$ defines a function $p = H(h)$, which is plotted in Fig. 1.

Problem 2.   Show that the trivial solution $\xi = \eta = \rho = 0$ is always unstable
(Murray, 1974).   Problem I.4 is a special case: $\epsilon = 0.1$, $p = 1$, $h = 7.2$.

Problem 3.   (a) For $h > 1/2$ and $0 < q \ll 1$, show that the stability
relation $\alpha\beta = C$ is approximated by

$$Tr = \Theta(\epsilon/p^2) , \quad \text{as} \quad \epsilon \to 0 ,$$

where

$$Tr = \frac{\beta\gamma - \alpha\delta}{\alpha} - \frac{1}{p} .$$

As long as $p \gg \sqrt{\epsilon}$, $Tr = 0$ is a good approximation of $\alpha\beta = C$. Show
that the steady state loses stability for

(8) $$p > p_c = \frac{(2h - 1)^2}{2(1 + 4h - 4h^2)} + \Theta(q) , \quad \text{as} \quad q \to 0 .$$

The critical value of p predicted by Eq. (8) agrees with that reported in
Fig. 1 to within .5% for .6 $< h <$ 1.2 .

  (b) For $h < 1/2$ and $0 < q \ll 1$, show that

(9) $$p_c = (1 + \frac{\epsilon}{q}) \frac{q}{4h - 1} , \quad \text{as} \quad \epsilon, q \to 0 \quad \text{such that} \quad \frac{\epsilon}{q} = \text{constant.}$$

Eq. (9) agrees with Fig. 1 to within 10% .

Fig. 1. The steady state (ξ_0, η_0, ρ_0) is unstable for $p > H(h)$.

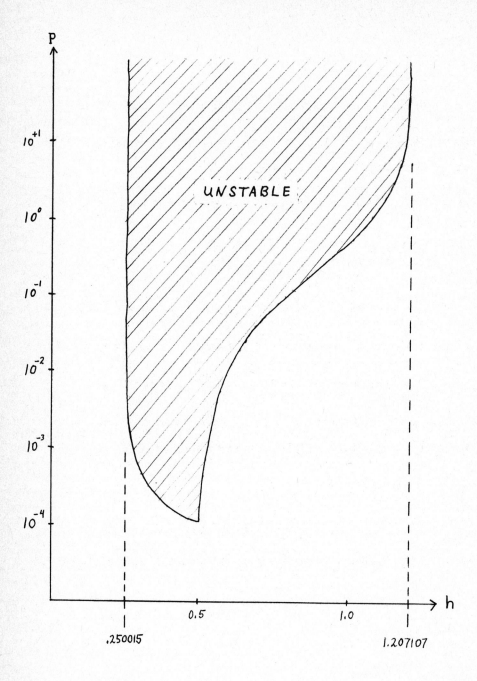

Existence of periodic solutions.

What happens when (ξ_0, η_0, ρ_0) is unstable?

First let us show that solutions of (1) are bounded (Murray, 1974a).

Consider the set

$$\mathbb{B} = \left\{ (\xi, \eta, \rho) \,\middle|\, 1 \le \xi \le \frac{1}{q}, \quad \frac{2hq}{1+q} \le \eta \le \frac{h}{q}, \quad 1 \le \rho \le \frac{1}{q} \right\} .$$

Obviously $(\xi_0, \eta_0, \rho_0) \in \mathbb{B}$. What does the vector field (1) look like on $\partial\mathbb{B}$, the surface of $\mathbb{B}$? On the face $\xi = 1$,

$$\epsilon\dot{\xi} = 1 - q > 0 \qquad \text{for} \quad 0 < q < 1 .$$

That is, solutions of (1) enter $\mathbb{B}$ through the face $\xi = 1$. On the face $\xi = 1/q$.

$$\epsilon\dot{\xi} = \eta(1 - \frac{1}{q}) < 0 \qquad \text{for} \quad 0 < q < 1 ,$$

so solutions enter $\mathbb{B}$ through this face as well. On the face $\eta = 2hq/1+q$,

$$\dot{\eta} = 2h\rho - \frac{2hq}{1+q}(1+\xi) \ge 2h - \frac{2hq}{1+q}(1+\frac{1}{q}) = 0 .$$

Thus solutions of (1) enter $\mathbb{B}$ through this face as well, except possibly at the point $(1/q, 2hq/(1+q), 1)$. However, at this point

$$\ddot{\eta} = -\dot{\eta} + 2h\dot{\rho} - \dot{\xi}\eta - \xi\dot{\eta}$$

$$= 2h \frac{1}{p}\left(\frac{1}{q} - 1\right) - \frac{2hq}{1+q}\frac{1}{\epsilon}\left(\frac{1}{q} + \frac{2hq}{1+q} - \frac{1}{q} - \frac{2h}{1+q}\right)$$

$$= \frac{2h}{pq}(1-q) + \left(\frac{2h}{1+q}\right)^2 \frac{q}{\epsilon}(1-q) > 0 \qquad \text{for} \quad 0 < q < 1 .$$

Thus, if $\eta(0) = 2hq/(1+q)$, then there exists a $t_1 > 0$ such that $\eta(t) > 2hq/$ for $t \in (0, t_1)$. So solutions of (1) cannot leave $\mathbb{B}$ through the face $\eta = 2hq/(1+q)$. Continuing in this manner, we find that any solution of (1) which starts in $\mathbb{B}$ or on $\partial\mathbb{B}$ at $t = 0$ must lie entirely within $\mathbb{B}$ for all $t > 0$.

Furthermore, any solution which starts in the positive octant ($\xi > 0$, $\eta > 0$, $\rho > 0$) eventually enters $\mathbb{B}$. For suppose $0 < \xi < 1$. Then $\epsilon\dot{\xi} = \xi(1 - q\xi) + \eta(1 - \xi) > 0$ for $0 < q < 1$, so ξ must increase until $\xi \geq 1$. Similarly, if $\xi > 1/q$, then $\epsilon\dot{\xi} < 0$ and ξ must decrease until $\xi \leq 1/q$. If $0 < \rho < 1$, then $\rho\dot{\rho} > \xi - 1$ and thus eventually we must have $\rho\dot{\rho} \geq 0$ because eventually $\xi \geq 1$. Thus eventually ρ must increase until $\rho \geq 1$. Continuing this argument proves the assertion.

Now divide $\mathbb{B}$ into eight smaller boxes, $\mathbb{B}_1$ to $\mathbb{B}_8$, (see Fig. 2)

$$\mathbb{B}_1 : 1 \leq \xi \leq \xi_0 \quad, \quad 2hq/(1+q) \leq \eta \leq \eta_0 \;, \; 1 \leq \rho \leq \rho_0$$

$$\mathbb{B}_2 : \xi_0 \leq \xi \leq 1/q \;, \; 2hq/(1+q) \leq \eta \leq \eta_0 \;, \; 1 \leq \rho \leq \rho_0$$

$$\mathbb{B}_3 : 1 \leq \xi \leq \xi_0 \quad, \quad \eta_0 \leq \eta \leq h/q \qquad, \; 1 \leq \rho \leq \rho_0$$

$$\mathbb{B}_4 : \xi_0 \leq \xi \leq 1/q \;, \; \eta_0 \leq \eta \leq h/q \qquad, \; 1 \leq \rho \leq \rho_0$$

$$\mathbb{B}_5 : 1 \leq \xi \leq \xi_0 \quad, \quad 2hq/(1+q) \leq \eta \leq \eta_0 \;, \; \rho_0 \leq \rho \leq 1/q$$

$$\mathbb{B}_6 : \xi_0 \leq \xi \leq 1/q \;, \; 2hq/(1+q) \leq \eta \leq \eta_0 \;, \; \rho_0 \leq \rho \leq 1/q$$

$$\mathbb{B}_7 : 1 \leq \xi \leq \xi_0 \quad, \quad \eta_0 \leq \eta \leq h/q \qquad, \; \rho_0 \leq \rho \leq 1/q$$

$$\mathbb{B}_8 : \xi_0 \leq \xi \leq 1/q \;, \; \eta_0 \leq \eta \leq h/q \qquad, \; \rho_0 \leq \rho \leq 1/q$$

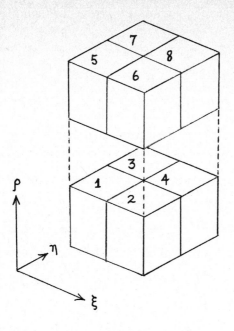

Fig. 2. Orientation of the boxes used by Hastings and Murray to prove the existence of oscillatory solutions to the Oregonator equations. The boxes have been separated on the $\rho = \rho_0$ plane to improve visibility.

Theorem (Hastings and Murray, 1975). <u>Assume</u> that the steady state (ξ_0, η_0, ρ_0) is unstable. Then with exactly two exceptions, solutions which enter $\mathbb{B}_4$ or $\mathbb{B}_5$ cannot remain in either of these boxes. Furthermore, no trajectory can enter either $\mathbb{B}_4$ or $\mathbb{B}_5$ from inside $\mathbb{B}$. A solution which enters one of the remaining boxes, either from outside $\mathbb{B}$ or from $\mathbb{B}_4$ or $\mathbb{B}_5$, must oscillate thereafter. Such a solution will eventually reach box $\mathbb{B}_3$, and thereafter proceed from box to box in the sequential order

$$\mathbb{B}_3 \to \mathbb{B}_1 \to \mathbb{B}_2 \to \mathbb{B}_6 \to \mathbb{B}_8 \to \mathbb{B}_7 \to \mathbb{B}_3 \to \ldots$$

Furthermore, there must exist at least one periodic solution which moves once through the boxes, as listed, before closing on itself.

Proof: (In spirit, this proof is similar to the two-dimensional example treated in Chapter I, pp. 17-20). Most of these results follow rather simply from examining the direction of the vector field on the faces of boxes 1 to 8 . For example, consider the "internal" faces of $\mathbb{B}_4$:

$$\xi = \xi_0 , \quad \eta \geq \eta_0 \Rightarrow \dot{\xi} \leq 0 \qquad \text{(leave } \mathbb{B}_4 \text{ to } \mathbb{B}_3 \text{)}$$

$$\eta = \eta_0 , \quad \xi \geq \xi_0 , \rho \leq \rho_0 \Rightarrow \dot{\eta} \leq 0 \qquad \text{(leave } \mathbb{B}_4 \text{ to } \mathbb{B}_2 \text{)}$$

$$\rho = \rho_0 , \quad \xi \geq \xi_0 \Rightarrow \dot{\rho} \geq 0 \qquad \text{(leave } \mathbb{B}_4 \text{ to } \mathbb{B}_8 \text{)} \quad .$$

Obviously, no trajectory can enter $\mathbb{B}_4$ from inside $\mathbb{B}$. Furthermore, since $\rho \leq \xi$, $\rho\dot{\rho} \geq 0$ and any trajectory must either tend to the steady state $(\xi_0 , \eta_0 , \rho_0)$ or intersect one of the three internal faces of $\mathbb{B}_4$ and enter $\mathbb{B}_2$, $\mathbb{B}_3$ or $\mathbb{B}_8$.

Suppose the trajectory enters $\mathbb{B}_2$. Again, since $\rho \leq \xi$, $\rho\dot{\rho} \geq 0$ and any trajectory must either tend to the steady state or intersect one of the three internal faces of $\mathbb{B}_2$. Examine the vector field on these faces:

$$\xi = \xi_0 , \quad \eta \leq \eta_0 \Rightarrow \dot{\xi} \geq 0 \qquad \text{(enter } \mathbb{B}_2 \text{ from } \mathbb{B}_1 \text{)}$$

$$\eta = \eta_0 , \quad \xi \geq \xi_0 , \rho \leq \rho_0 \Rightarrow \dot{\eta} \leq 0 \qquad \text{(enter } \mathbb{B}_2 \text{ from } \mathbb{B}_4 \text{)}$$

$$\rho = \rho_0 , \quad \xi \geq \xi_0 \Rightarrow \dot{\rho} \geq 0 \qquad \text{(leave } \mathbb{B}_2 \text{ to } \mathbb{B}_6 \text{)} \quad .$$

In a moment we shall prove that no trajectory can approach $(\xi_0 , \eta_0 , \rho_0)$ from within $\mathbb{B}_3 \cup \mathbb{B}_1 \cup \mathbb{B}_2 \cup \mathbb{B}_6 \cup \mathbb{B}_8 \cup \mathbb{B}_7$. Granted this fact, any trajectory which enters $\mathbb{B}_2$ must eventually intersect the face $\rho = \rho_0$ and enter $\mathbb{B}_6$

What happens to a trajectory which once enters $\mathbb{B}_6$? Examining the vector field on the interval faces, we find that the only exit from $\mathbb{B}_6$ is into $\mathbb{B}_8$. Thus a trajectory must either leave $\mathbb{B}_6$ and enter $\mathbb{B}_8$, or approach the steady state from within $\mathbb{B}_6$ (which we shall rule out momentarily), or it must wander around inside $\mathbb{B}_6$ forever. This last possibility can be ruled out, but the algebra gets messy. I leave the proof as an exercise, or consult Hastings and Murray (1975).

Continuing likewise with the other boxes, one proves that trajectories must circulate through the boxes $\mathbb{B}_3$, $\mathbb{B}_1$, $\mathbb{B}_2$, $\mathbb{B}_6$, $\mathbb{B}_8$, $\mathbb{B}_7$, $\mathbb{B}_3$ in the order listed, or else approach (ξ_0, η_0, ρ_0) from within one of these boxes. We must now eliminate this possibility. Close to (ξ_0, η_0, ρ_0) the solutions of (1) are well-approximated by solutions of the linearized system (5L)

$$
(5L) \qquad \dot{x} = Kx , \qquad x = \begin{pmatrix} \xi - \xi_0 \\ \eta - \eta_0 \\ \rho - \rho_0 \end{pmatrix} , \qquad K = \begin{pmatrix} -\alpha/\epsilon & -\beta/\epsilon & 0 \\ -\gamma & -\delta & 2h \\ 1/p & 0 & -1/p \end{pmatrix} .
$$

By assumption K has one real negative eigenvalue $(\lambda_1 < 0)$, and two eigenvalues with positive real part. Let $\mu = (\lambda_2 + \lambda_3)/2$. Obviously, $\mu > 0$. Only two solutions of (5L) approach $x = 0$ as $t \to +\infty$, namely

$$
x(t) = \pm e^{\lambda_1 t} \hat{x} , \qquad \hat{x} = \begin{pmatrix} \hat{x}_1 \\ \hat{x}_2 \\ \hat{x}_3 \end{pmatrix} ,
$$

where $\hat{x}$ is the eigenvector of K corresponding to the eigenvalue λ_1,

$$
(10) \qquad\qquad K\hat{x} = \lambda_1 \hat{x} .
$$

(Chapter I contains a discussion of solutions of systems of linear first order ordinary differential equations.) From (5L) and (10) we have

$$-\alpha \hat{x}_1 - \beta \hat{x}_2 = \epsilon \lambda_1 \hat{x}_1$$

$$\hat{x}_1 - \hat{x}_3 = p \lambda_1 \hat{x}_3$$

Since α, β, ϵ, p are all positive and $\lambda_1 < 0$, then $\hat{x}_2$ and $\hat{x}_3$ are nonzero if $\hat{x}_1$ is nonzero. Furthermore,

$$\lambda_1 + \lambda_2 + \lambda_3 = \lambda_1 + 2\mu = -\frac{\alpha}{\epsilon} - \delta - \frac{1}{p} < 0 \quad.$$

$$\epsilon \lambda_1 + \alpha = -\epsilon(2\mu + \delta + 1/p) < 0 \quad.$$

Thus, so $\hat{x}_2$ has the same sign as $\hat{x}_1$. As well, $p\lambda_1 + 1 = -p(2\mu + \delta + \alpha/\epsilon)$ < 0, so $\hat{x}_3$ has the opposite sign as $\hat{x}_1$.. Now a vector of sign pattern $(+, +, -)$ points from (ξ_0, η_0, ρ_0) into $\mathbb{B}_5$. This proves the assertion that two and only two trajectories approach (ξ_0, η_0, ρ_0) as $t \to +\infty$, one within $\mathbb{B}_4$ and the other within $\mathbb{B}_5$.

Finally, the theorem asserts the existence of at least one periodic solution. After a bit of algebra, this follows directly from the Brouwer fixed point theorem (see Hastings and Murray).

Limit cycles in the relaxation-oscillator regime.

Now we would like to determine some properties of the periodic solution, whose existence we have just discussed (Tyson, 1975). To develop expressions for the amplitude, period and wave form of the oscillations we will exploit the fact that ϵ and q are small quantities, whereas p is large (see p. 42). It is most convenient to consider the Oregonator as expressed in Eq. (5)

$$\epsilon \dot{x} = -\alpha x - \beta y - qx^2 - xy$$

(5)
$$\dot{y} = -\gamma x - \delta y + 2hz - xy$$

$$p\dot{z} = x - z$$

In the limit as $\epsilon \to 0$, $|\dot{x}| \to \infty$ unless the right hand side of Eq. (5a) is identically zero. That is, $x(t)$ will change very rapidly in order to maintain

$$-\alpha x - \beta y - qx^2 - xy = 0 \quad .$$

This defines x as a function of y

(11)
$$x = X(y) = \left[-(\alpha + y) + \sqrt{(\alpha + y)^2 - 4q\beta y} \right] (2q)^{-1} \quad .$$

(We take only the positive branch of the square root; the negative branch lies in the physically unacceptable region of negative concentrations.) Eq. (11) is said to define the "slow manifold", $\mathbb{m} = \{ (x,y,z) \, | \, x = X(y) \}$. At fixed y, if $x > X(y)$, then $\epsilon \dot{x} < 0$ and, since ϵ is small, x will decrease quickly. On the other hand, if $x < X(y)$, then $\epsilon \dot{x} > 0$ and x will increase quickly. That is, any initial conditions $(x(0), y(0), z(0)) \notin \mathbb{m}$ will evolve so that at time $\tau_1 \sim \mathcal{O}(\epsilon)$: $x(\tau_1) \sim X(y)$, $y(\tau_1) \sim y(0)$ and $z(\tau_1) \sim z(0)$. Thereafter, i.e. for $\tau > \tau_1$, $(x(\tau), y(\tau), z(\tau))$ will remain close to $\mathbb{m}$. System (5) reduces essentially to the two dimensional system.

(12)
$$\dot{y} = -\gamma X(y) - \delta y + 2hz - X(y)y$$
$$p\dot{z} = X(y) - z \quad .$$

The slow manifold, $\mathbb{m}$, is illustrated in Fig. 3.

Let us derive some properties of the slow manifold for $0 < q \ll 1$. We will restrict ourselves to the case $h > 1/2$. Similar equations for $h < 1/2$ can be found in Tyson (1975). First, improve approximation (4c)

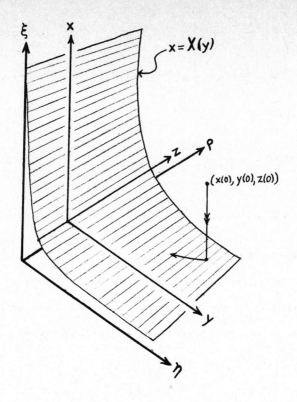

Fig. 3.   The slow manifold, $\mathbb{m} = \{(x,y,z) \,|\, x = X(y),$ given by Eq. (11)$\}$, for
q small. From arbitrary initial data $(x(0)\,,\,y(0)\,,\,z(0))$ solutions
of system (5) evolve quickly to the slow manifold and stay on $\mathbb{m}$
thereafter.

(4c)

$$\rho_0 = \xi_0 = \frac{2h+1}{2h-1} - 4h \, \frac{2h+1}{(2h-1)^3} \, q + \mathbb{O}(q^2) \left.\begin{array}{c} \\ \\ \end{array}\right\}$$

$$\eta_0 = \frac{2h+1}{2} - \frac{2h+1}{2(2h-1)} \, q + \mathbb{O}(q^2) \qquad \text{for} \quad h > 1/2 \quad .$$

Then, to first order in q ,

$$\alpha = \frac{2h-1}{2} + \frac{2h+1}{2(2h-1)}\, q \ , \qquad \beta = \frac{2}{2h-1} - 4h\,\frac{2h+1}{2(2h-1)^3}\, q$$

(13)

$$\gamma = \frac{2h+1}{2} - \frac{2h+1}{2(2h-1)}\, q \ , \qquad \delta = \frac{4h}{2h-1} - 4h\,\frac{2h+1}{2(2h-1)^3}\, q \ .$$

Furthermore,

(11′) $X(y) = \begin{cases} -\dfrac{\beta y}{\alpha+y} - \left(\dfrac{\beta y}{\alpha+y}\right)^2 \dfrac{q}{\alpha+y} + \mathbb{O}(q^2) & , \quad \text{for} \quad \alpha+y > 0 \\[4mm] -\dfrac{\alpha+y}{q} + \dfrac{\beta y}{\alpha+y} + \left(\dfrac{\beta y}{\alpha+y}\right)^2 \dfrac{q}{\alpha+y} + \mathbb{O}(q^2) & , \quad \text{for} \quad \alpha+y < 0 \ . \end{cases}$

Remember that $\eta = \gamma + y > 0$.

Now consider the planar system (12) for $p \gg 1$. In this case y changes very much faster than z ($\tau \sim 1$ as compared to $\tau \sim p$). Again we can argue that, before z changes appreciably, y will change so that

$$-\gamma X(y) - \delta y + 2hz - X(y)y = 0 \ .$$

This defines a second slow manifold $\hbar = \{(x,y,z)\,|\,x = X(y), y = Y(z)\}$ where $Y(z)$ is defined implicitly by

(14)
$$2hz = \delta y + \gamma X(y) + yX(y) \ .$$

Notice that $\hbar \subset \hbar$.

What does $\hbar$ look like for $0 < q \ll 1$, $h > 1/2$? From Eqs. (11′), (13) and (14) we derive:

(15) $z \approx \dfrac{1}{h}\, y - \dfrac{2h+1}{2h(2h-1)}$ \hfill for $y \gg 1$

(16) $z \approx \dfrac{4}{h(2h-1)^3}\left[y + \dfrac{1}{8}\,(4h^2 - 4h - 1)(2h-1)\right]^2 - \dfrac{(4h^2 - 4h - 1)^2}{16h(2h-1)}$, \hfill for $y \approx 0$

$$(17) \quad z \approx \frac{1}{8hq} - \frac{1}{2hq}(y+h)^2 - \frac{8(2h+1)y^2 + 2(8h^2 - 2h + 1)y - (4h^2 - 1)}{4h(2h-1)(1-2h-2y)} \quad ,$$

for $y < -h + 1/2$.

Notice that for $y = -\eta_0$, Eq. (17) gives $z = -\rho_0 + \mathcal{O}(q)$. From Eqs. (15) - (17) we can patch together a reasonably accurate picture of Eq. (14), as in Fig. 4. We require that $4h^2 - 4h - 1 < 0$, i.e. $h < (1 + \sqrt{2})/2$, so that point B lies in the fourth quadrant.

For $p \gg 1$, y changes much more quickly than z . Starting at initial conditions P , the solution of (12) moves quickly (time ≈ 1) along a horizontal line ($\Delta z = 0$) to R where $\dot{y} = 0$. Since $\dot{z} < 0$ at point R , the trajectory now proceeds slowly (time $\approx p$) along the line AB . At point B the trajectory leaves the $\dot{y} = 0$ nullcline and jumps (time ≈ 1) to point C where again $\dot{y} = 0$. Along CD , $\dot{z} > 0$ so z increases until at point D the trajectory jumps to A . Since $\dot{z} < 0$ at A , z decreases slowly along section AB and the cycle repeats itself.

From Eqs. (15) - (17) we easily derive expressions for the maximum and minimum values of y and z around the cycle.

$$\text{At D} \begin{cases} y_{\text{jump up}} = -h + \mathcal{O}(q) \\[2mm] z_{\text{max}} = \dfrac{1}{8hq} - \dfrac{8h^2 - 2h + 1}{4h(2h-1)} + \mathcal{O}(q) \end{cases}$$

$$\text{At A} \begin{cases} z = z_{\text{max}} \\[2mm] y_{\text{max}} = \dfrac{1}{8q} - \dfrac{16h^2 - 10h - 1}{4(2h-1)} + \mathcal{O}(q) \end{cases}$$

$$\text{At B} \begin{cases} y_{\text{jump down}} = -\dfrac{1}{8}(4h^2 - 4h - 1)(2h-1) + \mathcal{O}(q) \\[2mm] z_{\text{min}} = -\dfrac{(4h^2 - 4h - 1)^2}{16h(2h-1)} + \mathcal{O}(q) \quad . \end{cases}$$

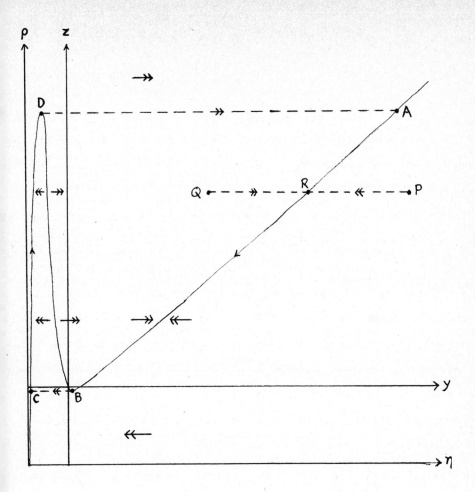

Fig. 4. Schematic diagram of the $\dot{y} = 0$ nullcline, Eq. (14), for
$0.5 < h < 1.207$. Points A , B , C , D correspond to the
characteristic points of the oscillation recorded in Fig. 1.
Analytic expressions for these points are given in the text.

$$\text{At C} \begin{cases} z = z_{min} \\ \\ y_{min} = -h - \dfrac{1}{2} - \dfrac{(4h^2 - 4h - 1)^2 - 8h(2h+1)(4h+1)}{8(2h-1)} q + \Theta(q^2) \end{cases}$$

We can also derive approximate expressions for the period along phases AB and CD :

$$T_{AB} = \int_{z_{max}}^{z_{min}} dz/\dot{z} = \int_{z_{max}}^{z_{min}} pdz/[X(y) - z]$$

$$\approx \int_{z_{max}}^{z_{min}} \frac{-pdz}{z + \dfrac{2}{2h-1}} = p \, \ell n \left[\frac{1}{8q} \, \mathcal{T}_{AB}(h) \right]$$

where

$$\mathcal{T}_{AB}(h) = \frac{2h - 1}{2h - (h^2 - h - \frac{1}{4})^2} + \Theta(q) \ .$$

Problem 4. Show that

$$T_{CD} \approx p \, \mathcal{T}_{CD}(h) \ ,$$

where

$$\mathcal{T}_{CD}(h) = \frac{4h - 1}{2h - 1} \, \ell n \left[\frac{4h}{4h - 1} \left(\frac{1}{2} \right)^{\frac{1}{4h-1}} \right] \ .$$

Show that

$$\lim_{h \to \frac{1}{2}} \mathcal{T}_{CD}(h) = 2 \, \ell n \, 2 - 1 \approx 0.4 \ .$$

A few calculations show that $\mathcal{T}_{CD}(h)$ is monotonically decreasing for $h > \frac{1}{2}$

A characteristic feature of the observed oscillation (see Fig. II.1) is exponential decay of $[Br^-]$ during the slow Br^- consumption phase AB .

From our analysis we have

$$p\dot{z} = X(y) - z$$

along section AB , or

$$\dot{z}(\tau) \sim \exp\left(-\frac{1}{p}\,\tau\right) \quad.$$

Since $y \sim hz$ along section AB ,

$$\ln y = -\frac{\tau}{p} + \text{constant} \quad.$$

Since $\tau = 0.1 \text{ sec}^{-1} t$ and $y \sim \text{const.} \times [Br^-]$,

$$\log [Br^-] \approx -\frac{0.1 \text{ sec}^{-1}}{2.3\,p}\, t$$

along section AB . From Fig. II.1, $\log [Br^-] \approx -(.02 \text{ sec}^{-1})t$, thus $p = 2$.

This value of p is much smaller than our original estimate $(p \sim 300)$, which simply means that the Oregonator is not quantitatively correct, if we insist on the values of the parameters suggested by experiment. This is borne out further by considering $q \sim 10^{-5}$. In this case

$$[Br^-]_{max} \sim (3.3 \times 10^{-7} \text{ M})y_{max} \sim \frac{3.3 \times 10^{-7} \text{ M}}{8 \times 10^{-5}} \sim 4 \times 10^{-3}$$

or $\log [Br^-]_{max} \sim -2.4$, which is much too large. To fit the observed $\log [Br^-]_{max} \sim -5$, we must choose $q \sim 4 \times 10^{-3}$.* Parameters h and ϵ are harder to pin down, because the period and amplitude of oscillation are not sensitive to these parameters. But for just this reason, the choice of h and ϵ is not critical. Since we must choose q considerably larger than originally anticipated, it seems reasonable to choose ϵ larger as well.

---

*It should be mentioned that calculations on a more complete model of the FKN mechanism (Edelson, Field and Noyes, 1975) reduce this discrepancy considerably: the amplitude comes out correct and the period is only three times too long. Better knowledge of the kinetics of cerium oxidation of organic compounds may correct the residual error in period.

In Fig. 5 we compare the limit cycle solution of Eq. (5) for $\epsilon = .03$, $p = 2$, $q = .006$, $h = .75$ with the observations previously cited in Fig. 1? The agreement seems fair enough considering the drastic assumptions made in deriving the Oregonator equations.

In Fig. 6 we compare the calculated limit cycle with the analytic representation derived in this section. As p gets larger, the approximations get better and better.

Hard self-excitation.

Consider the linearization of Eq. (12), using Eq. (11′) for $y \approx 0$,

$$\dot{y} = - \frac{\alpha\delta - \beta\gamma}{\alpha} y + 2hz$$

(12L)

$$p\dot{z} = -(\beta/\alpha)y - z \quad .$$

The characteristic equation (see Chapter I) for this planar linear system is

$$\lambda^2 - \text{Tr}\,\lambda + \text{Det} = 0$$

where

$$\text{Tr} = - \frac{\alpha\delta - \beta\gamma}{\alpha} - \frac{1}{p}$$

$$\text{Det} = (\alpha\delta - \beta\gamma + 2h\beta)/\alpha p \quad .$$

Using Eq. (13),

(18)
$$\left.\begin{array}{l} \text{Det} = \dfrac{2}{p} \cdot \dfrac{2h+1}{2h-1} > 0 \\[3mm] \text{Tr} = - \dfrac{2(4h^2 - 4h - 1)}{(2h-1)^2} - \dfrac{1}{p} \end{array}\right\} \quad \text{for} \quad h > 1/2 \ .$$

The steady state $(0,0)$ is stable if and only if $\text{Tr} < 0$, i.e.

Fig. 5. Comparison of observed oscillation (Field, Körös, Noyes, 1972, Fig. 5),
         solid line, with limit cycle solution of Eq. (5) for $\epsilon = .03$, $p = 2$,
         $q = .006$, $h = .75$, dashed line. To convert to real concentration
         and time, we have used $[Br^-] = (y + 1.25)/3 \times 10^6 M^{-1}$, $t = (12 \sec)\tau$.

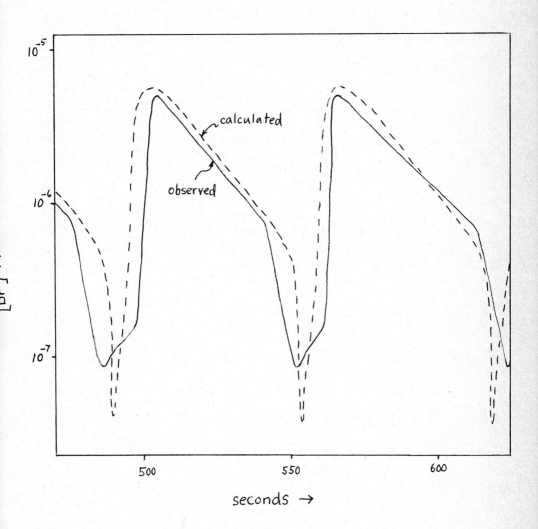

Fig. 6. Limit cycle calculated from Eq. (5) for ϵ = .03 , q = .006 ,
h = .75 : p = 2 (heavy line) and p = 10 (dotted line); compared
to analytic representation   ABCDA   (light line and dashed lines).

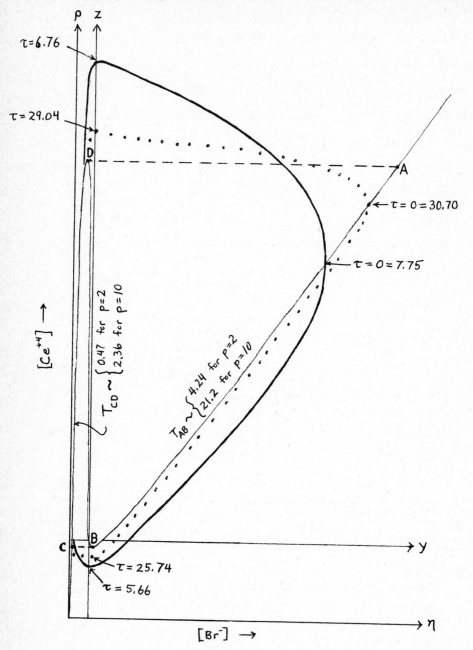

(19)
$$p < p_c = \frac{(2h - 1)^2}{2(1 + 4h - 4h^2)} , \quad \text{for} \quad h > 1/2 .$$

(See Problem 3.)

Notice that $p_c \to +\infty$ as $h \to (1 + \sqrt{2})/2 \approx 1.207$ from below. For $h > 1.207$ the steady state is locally stable for all values of p. But we can say much more about solutions of Eq. (12) for $h > 1.207$. From the expressions for y and z at point B (see p. 57) we see that point B lies in the third quadrant ($y < 0$, $z < 0$) for $h > 1.207$. In this case, when p is large, there no longer exists a periodic solution because solutions proceed directly to the origin, as illustrated in Fig. 7. Furthermore, from any initial conditions (e.g. P, Q, R, S in Fig. 7) the solution of Eq. (12) eventually approaches the steady state.

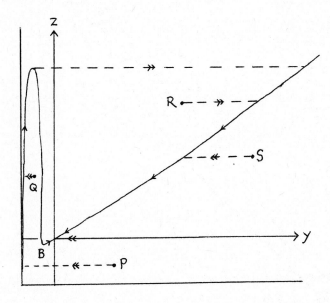

Fig. 7. Phase plane for $h > 1.207$. When p is large, the steady state, $y = 0$, $z = 0$, is globally asymptotically stable.

We also see from Fig. 7 that the steady state is a node for $h > 1.207$ and p large. The line of demarcation between nodal and spiral behavior close to the steady state is given precisely by $\mathrm{Tr}^2 = 4\,\mathrm{Det}$. Using Eqs. (18) this becomes

(20)
$$\frac{4(4h^2 - 4h - 1)^2}{(2h - 1)^4}\, p^2 - \frac{4(4h^2 + 4h - 1)}{(2h - 1)^2}\, p + 1 = 0 \quad .$$

The roots of Eq. (20) are plotted in Fig. 8 along with p_c , as given by Eq. (19).

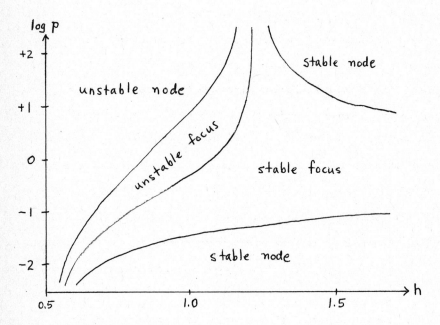

Fig. 8. Character of trajectories close to the steady state.

For h slightly less than 1.207 , p_c is a large number. Thus, it is possible to choose p large enough (say, $p > 5$) such that the stable limit cycle discussed in the previous section exists and yet small enough

$(p < p_c)$ such that the steady state is locally stable. In this case there must exist an unstable limit cycle surrounding the origin inside the stable limit cycle. The unstable limit cycle divides the y, z plane into two parts: a domain of attraction of the stable steady state and a domain of attraction of the stable limit cycle. Fig. 9 illustrates the situation for $p = 4.6$ and $p = 24$ with $h = 1.17$, $q = .006$, $\epsilon = .03$.

This phenomenon is often called "hard self-excitation" because there exists a self-excited (i.e. orbitally asymptotically stable) limit cycle, but to reach the self-excited oscillation requires a "hard" (i.e. finite) perturbation from the steady state. (In contrast, a "soft self-excitation" is illustrated in Fig. I.1.) There is some experimental indication of hard self-excitation in the Belousov-Zhabotinskii reaction. Notice in Fig. II.1 that after a short induction period the oscillations appear suddenly with large amplitude. This is to be expected for hard self-excitation: during the induction period the system is trapped in a locally stable steady state until the kinetic parameters change such that the steady state loses its stability and the system jumps to large amplitude stable oscillations. In the case of soft self-excitation it is expected that as the steady state loses stability, small amplitude stable oscillations first appear and then grow in size.

We have seen that hard self-excitation is to be expected for h slightly less than 1.207. What can we expect over the whole range $0.5 < h < 1.207$? To answer this question requires a rather delicate analysis of the nonlinear terms in Eq. (12) when $p = p_c$ (Tyson, 1975). The conclusion is that system (12) exhibits hard self-excitation for all

Fig. 9. Stable (———) and unstable (-----) limit cycles for p = 4.6 and p = 24. The other parameters are: h = 1.17, q = .006, ε = .03, in which case p_c = 24.276. At p = 24 the unstable limit cycle is small and harmonic. As p decreases, the unstable limit cycle grows larger and more "nonlinear." At p = 4.4 both limit cycles have disappeared and the origin is globally asymptotically stable.

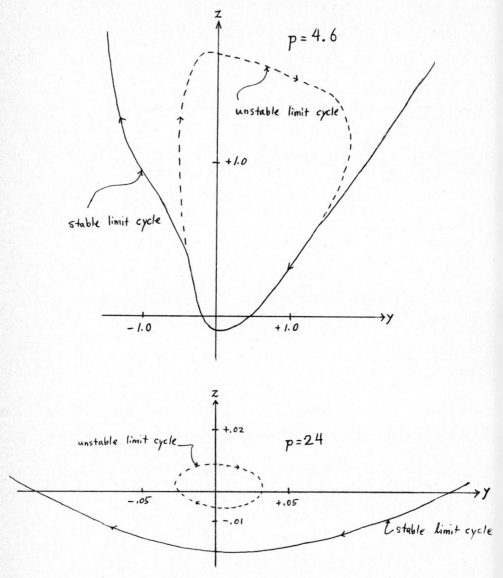

$h \in (0.5, 1.207)$ and for p sufficiently close to p_c, i.e. $-1 \ll p - p_c < 0$.

We have been primarily concerned with the two-dimensional version (12) of the full Oregonator equations (5). Stanshine (1975) has derived a rather complicated asymptotic representation of the limit cycle solutions of (5) in a limit equivalent to $\epsilon \to 0$, with $q = \mathcal{O}(\epsilon^{3/2})$ and $p = \mathcal{O}(\epsilon^{1/2})$. Troy and Field (1975) have some results on global asymptotic stability of (ξ_0, η_0, ρ_0) in the same limit. Hsü and Kazarinoff (1975) have investigated the appearance of hard self-excitation for system (5).

A different three-dimensional model of the BZ reaction has been suggested by Zhabotinskii, Zaikin, Korzukhin and Kreitser (1971). It is discussed in an appendix.

CHAPTER IV.   CHEMICAL WAVES

Waves of chemical activity in unstirred Belousov-Zhabotinskii reagent can be conveniently classified as

1.   kinematic waves

  a.   phase gradient waves

  b.   frequency gradient waves

2.   trigger waves

  a.   pacemaker waves

  b.   scroll waves

(see Winfree, 1974c).

Kinematic waves (first reported by Busse, 1969) only appear in self-oscillatory reagent and merely expose local phase or frequency variations which inevitably develop in unstirred medium of large enough extent.   Such waves are independent of diffusion, since they are not stopped by impermeable barriers (Kopell and Howard, 1973a).

On the other hand, trigger waves (first reported by Zaikin and Zhabotinsky, 1970) may appear in quiescent as well as oscillatory reagent and are dependent on diffusion.   They are waves of excitation conducted through the medium in a manner analogous to the spread of a grass fire.

Both kinds of waves are easily produced experimentally.   Kinematic waves can be observed using the same recipe for oscillations given at the beginning of Chapter III.   To produce spatial inhomogeneity, dissolve the malonic acid and sodium bromate in sulphuric acid along with a few mls of Ferroin in the bottom of an ungraduated cylinder.   In a separate beaker dissolve the cerium salt in about 100 ml of water.   Then carefully pipette the cerium solution on top of the sulphuric acid solution (it will float).   With a glass rod mix the

two solutions together a little bit. Blue bands will form near the interface
and move up or down, depending presumably on the exact nature of the induced
inhomogeneity. As the solutions are mixed more thoroughly, bands will form
throughout the cylinder. Most often bands are observed forming near the bottom
of the cylinder and moving upwards. Succeeding bands, emitted at the bottom,
move more and more slowly so that after a few minutes the cylinder is full of
bands packed closely at the bottom and spaced out towards the top.

Winfree (1972) has reported convenient conditions for observing target
patterns and scroll waves.

1. Dissolve 3 ml concentrated sulphuric acid and 10 g sodium bromate in
    134 ml water.

2. Dissolve 1 g sodium bromide in 10 ml water.

3. Dissolve 2 g malonic acid in 20 ml water.

In a small glass beaker add ½ ml of solution 2 to 6 ml of solution 1. Then
add 1 ml of solution 3 and wait a few minutes for the solution to become
clear. Then add 1 ml of .025M (standard) Ferroin. Mix well, pour into a
90 mm petri dish and cover it. The solution is uniformly red, but in a few
minutes blue dots will appear and spread out in rings. Shortly the dish will
be full of target patterns. Spiral waves can be produced by gently tipping
the dish so as to break some of the blue wave fronts. Free ends wrap around
into spiral structures.

Kinematic waves.

(a) Phase gradients

Suppose we have oscillatory reagent in a tall cylinder and we have
arranged that the period is everywhere the same $(= T)$ but that the phase
of oscillation varies linearly from top to bottom. By this we understand

that the state of the system can be represented by a 2π-periodic function of phase, ϕ . In particular, we associate the sharp, leading edge of a blue wave with $\phi = 2\pi n$, $n = 0,\pm 1,\pm 2,\ldots$

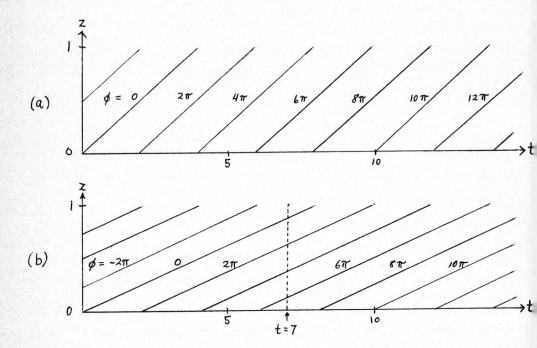

Fig. 1. Curves of constant phase, $\phi(z,t) = 2\pi n$, for a cylinder, $0 < z < 1$, with an initial phase gradient but no frequency gradient.

In Fig. 1a, $T = 2$ and the phase varies through two cycles from top to bottom, the bottom being ahead of the top. At $t = 0$ a blue wave appears at the bottom, travels up the tube at constant velocity, $v = 1/4$, and disappears at the top at $t = 4$. Meanwhile at $t = 2$ a second wave appears at the bottom and follows the first at the same velocity, and so forth. Obviously, if we were to insert an impermeable barrier at some height z , $0 < z < 1$, it would have no effect on the wave. In Fig. 1b the phase gradient is steeper, varying through four cycles, and the velocity is correspondingly smaller, $v = 1/8$. At $t = 7$ there are four evenly spaced bands in the cylinder, all moving upwards at the same velocity.

In general the velocity is inversely proportional to the phase gradient. For, if there is no frequency gradient, then

$$\phi(z,t) = \phi_0(z) + 2\pi t/T$$

and the velocity of a wave front, $\phi = 2\pi n$, is

$$v = \frac{dz}{dt}\bigg|_{\phi = const} = -\frac{\partial\phi/\partial t}{\partial\phi/\partial z} = -\frac{2\pi}{T}\frac{1}{\partial\phi_0/\partial z} \ .$$

In Fig. 1 $\partial\phi_0/\partial z < 0$, $v > 0$, i.e. bands move upward.

The experiment considered here would be rather difficult to arrange but Winfree (1974c) reports conditions under which phase gradients, in the absence of frequency gradients, naturally arise.

(b) Frequency gradients

It is easier to arrange frequency gradients as suggested earlier by layering aqueous solution on top of sulphuric acid solution, or better yet by applying a temperature gradient to an otherwise homogeneous solution in a cylinder (Kopell and Howard, 1973a). The temperature gradient induces a frequency gradient. One observes kinematic waves which are packed ever closer together because, as time progresses, phase gradients steepen and velocities decrease.[*]

For example, consider Fig. 2. The period T is a monotonic increasing function of height, z; in fact, $T(z) = 1+z$. At $t = 0$ the entire cylinder turns blue, that is, phase is initially zero throughout the cylinder. At $t = 1$ a blue band appears at the bottom and moves upward at velocity $v = 1$.

---

[*] Simultaneously and independently Theones (1973) published the same explanation of frequency gradient waves. Unfortunately, he erred in claiming to explain trigger waves (target patterns and spirals) in terms of frequency gradients without diffusive coupling between spatially neighboring points.

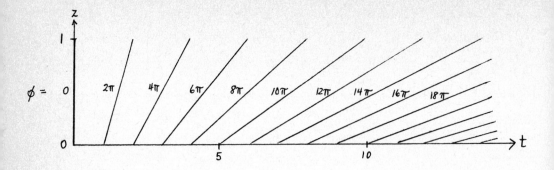

Fig. 2. Curves of constant phase, $\phi(z,t) = 2\pi n$, for a cylinder, $0 < z < 1$, with a frequency gradient but no initial phase gradient.

At $t = 2$ a second band appears and moves upward at $v = 1/2$. At $t = k$ the kth band appears and moves upward at $v = 1/k$.

Problem 1. For the example in Fig. 2 show that at $t = 2k + \frac{1}{2}$ there are k bands in the cylinder at positions

$$z_i = \frac{2i+1}{2(2k-i)} \quad , \qquad i = 0,\ldots,k-1 \ .$$

In general we can write

$$\phi(z,t) = \phi_0(z) + \psi(z,t)$$

where $\phi_0(z)$ is the initial phase distribution and $\psi(z,t)$ satisfies

$$\psi(z,0) = 0 \ , \ \psi(z, t+T(z)) = \psi(z,t) + 2\pi$$

where $T(z)$ is the local autonomous period. For simplicity, suppose $\phi_0(z) = 0$

Let $t_k(z)$ satisfy

$$\phi(z, t_k(z)) = 2\pi k ,$$

i.e. $t_k(z)$ is the time at which the kth wave front passes position z .
Calling the initial synchronous oscillation the $k = 0$ wave, we have

$$t_k(z) = kT(z)$$

and the velocity of the kth wave is

$$v_k = \left.\frac{dz}{dt}\right|_{\phi = 2\pi k} = \left(\frac{dt_k(z)}{dz}\right)^1 = \frac{1}{kT'(z)} .$$

If $T(z)$ is monotone increasing, $T' > 0$, then the kth blue wave enters
at $z = 0$, $t = kT(0)$, propagates up the tube at $1/k$ times the velocity of
the first wave, and leaves at $z = 1$, $t = kT(1)$. Since the velocity decreases as k increases, the waves are packed ever closer together at the
bottom of the tube.

Eventually diffusion effects will become important at the bottom of the
tube. Using our previous estimate of diffusion constants for small molecules
($D \approx 10^{-5} cm^2/sec$, see p. 21), we find that over a period $T(0) \approx 50$ sec
molecules diffuse a distance

$$d = \sqrt{2Dt} \approx .03 \, cm .$$

For what k are the bands this close together? The velocity of the first
wave is $v_1 \approx 1$ inch/min $\approx .04 \, cm/sec$, so the kth wave has travelled
$v_k T(0) = 2 \, cm/k$ by the time that the $k + 1$ th wave is emitted. We do not
have $.03 \approx 2/k$ until $k \approx 70$.

Problem 2. What is the band pattern like for a monotonically decreasing period gradient, such as

$$T(z) = \frac{1}{1-z} \quad , \quad 0 < z < 1 \ ?$$

Trigger waves.

Observationally, there are two classes of waves which depend for their existence on the interaction of reaction and diffusion. Target patterns (axisymmetric periodic travelling waves or, if you prefer, "bull's-eyes") are distinguished by their symmetry and by the fact that frequency and wavelength vary from pattern to pattern. Furthermore, target patterns do not appear in Z reagent which is carefully filtered and observed in a new petri dish coated with silicone resin (Winfree, 1974b, c). After deliberate contamination with dust, target patterns reappear. These facts clearly suggest that target patterns are organized about a heterogeneous nucleus, which serves as pacemaker for the pattern. Suppose that near a particle of dust or scratch in the glass the physico-chemical conditions are such as to produce a local oscillation of period $T < T_0$, where T_0 is the period of oscillation of the bulk medium. (T_0 may be infinite, as in Winfree's non-oscillatory Z reagent.) A wave of excitation will then be initiated at the heterogeneous nucleus every T seconds and spread radially through the excitable medium at a velocity which depends on properties of the bulk medium alone. (We shall discuss wave propagation shortly.) Since T presumably depends on special characteristics of the heterogeneneity, the wave spacing should vary from pattern to pattern, being directly proportional to T .

Scroll waves, on the other hand, are observed in the carefully filtered reagent given proper initial conditions (Winfree, 1974b, c). They come in several varieties (symmetric spirals, elongated spirals, elongated rings, scroll rings), but every such pattern in a given medium has the same rotational frequency and wave spacing. It is unlikely that spatial inhomogeneities play any significant role in the generation of scroll waves.

Scroll waves and target patterns (as well as solitary trigger waves) share the common property of dependence on the interaction of reaction and diffusion, as witnessed by the fact that they are all blocked by impermeable barriers.

A distinctive feature of the reaction is its excitability, which we shall illustrate by reference to the two-dimensional model (see Chapter III):

$$\dot{y} = -\delta y + 2hz - (\gamma + y)X(y) \qquad , \qquad y \sim [Br^-] - [Br^-]_0$$

$$(1) \qquad p\dot{z} = X(y) - z \qquad , \qquad z \sim [Ce^{+4}] - [Ce^{+4}]_0$$

$$X(y) = \left[-(\alpha + y) + \sqrt{(\alpha + y)^2 - 4q\beta y} \right](2q)^{-1} \quad , \qquad X \sim [HBrO_2] - [HBrO_2]_0$$

for $h > 1.207$ and p large. In this case the steady state $(y = 0, z = 0)$ is globally asymptotically stable; however, in response to certain finite disturbances of y and z the system goes off on a long excursion before returning to the vicinity of the steady state. See Fig. 3.

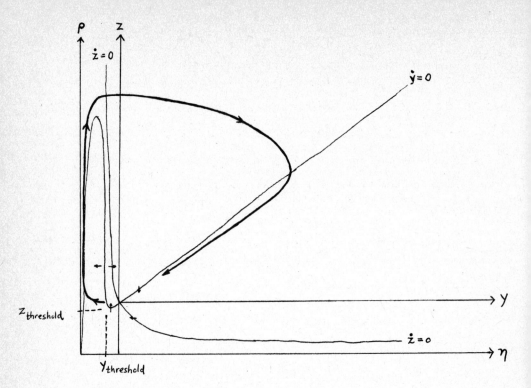

Fig. 3. Phase plane for system (1) with $h > 1.207$ and p large. Plotted are the $\dot{y} = 0$ and $\dot{z} = 0$ nullclines and a representative trajectory started at $y = y_{threshold}$, $z = 0$.

The threshold values of y and z, as indicated on Fig. 3, are easily derived from the relations given in the last chapter (p. 57):

$$y_{threshold} = -(4h^2 - 4h - 1)(2h - 1)/4 < 0 , \quad \text{for} \quad h > 1.207$$

$$z_{threshold} = -\frac{(4h^2 - 4h - 1)^2}{16h(2h - 1)} < 0 .$$

The resting state $y = 0$, $z = 0$ corresponds to low $[Ce^{+4}]$ for $h > 1/2$:

$$[Ce^{+4}]_0 \sim (1.7 \times 10^{-4} M) \frac{2h + 1}{2h - 1} .$$

A disturbance from the resting state that reduces y below $y_{threshold}$ and/or z below $z_{threshold}$ results in a transient excursion, during which time z increases above $z_{max} \sim 1/8hq$ (see p. 57). That is, the $[Ce^{+4}]$ increases dramatically, which corresponds to the blue dot appearing in the red medium. During this transient response $[Br^-]$ decreases dramatically at the point of the initial disturbance. But this will cause y to drop below $y_{threshold}$ in neighboring volume elements, and so the disturbance propagates through the medium.

Problem 3. Show that, if $h < 1/2$, then model (1) predicts a blue resting state (large $[Ce^{+4}]$) through which a red wave propagates.

Troy and Field (1975) have proved similar results on excitability for the full three-dimensional Oregonator equations (III.1).

From the time course predicted by Fig. 3 and the observation that waves propagate at approximately constant velocity we can sketch roughly the waveforms of $[Br^-]$, $[Ce^{+4}]$ and $[HBrO_2]$ in a trigger wave, as in Fig. 4, which should be compared with the concentration profiles reported by Field and Noyes (1972). At the leading edge of the wave, $[Br^-]$ drops and $[HBrO_2]$ increases dramatically. Diffusion will cause $[Br^-]$ to decrease and $HBrO_2$ to increase ahead of the wave, triggering a transient excitation in neighboring volume elements. Behind the wave, $[Br^-]$ is very large, having been regenerated by the oxidation of bromomalonic acid by Ce^{+4} . The large $[Br^-]$ makes the system "refractory" just behind the propagating blue wave. The medium cannot be re-excited until $[Br^-]$ and $[Ce^{+4}]$ return to the vicinity of the steady state. This refractionness explains the mutual annihilation of trigger waves upon collision.

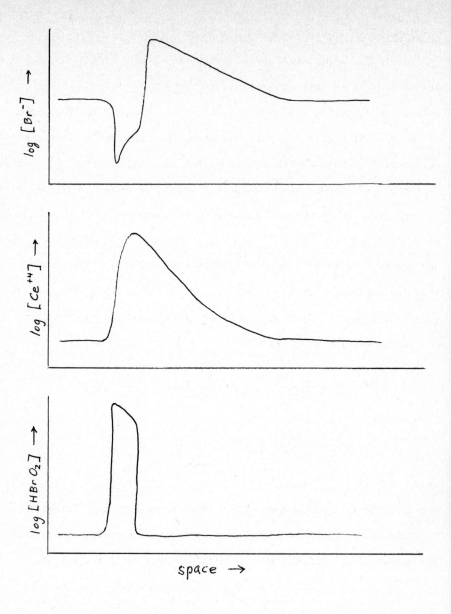

Fig. 4. Approximate concentration profiles of Br^-, Ce^{+4} and $HBrO_2$ in a trigger wave (single pulse) moving from right to left.

Velocity of propagation of trigger waves.

Field and Noyes (1974b) have suggested a simple model for the propagation of trigger waves based on the FKN mechanism (Chapter II). In their model, at a wave front the reaction is switching from process (A) to process (B): $HBrO_2$ increases dramatically from front to rear and Br^- decreases. (Behind the wavefront process (C) regenerates Br^-, which in turn destroys $HBrO_2$ by reaction (R2), and the medium eventually returns to its initial condition.) The wavefront propagates by the diffusion of $HBrO_2$ forward and Br^- backward, which triggers the switch from process (A) to process (B) in the volume element just ahead of the wave. Ignoring the regeneration of Br^- by Ce^{+4}, they suggest that steps (M1) - (M4) of the Oregonator (Chapter III) provide a reasonable description of the propagating wave front:

$$
\frac{\partial X}{\partial t} = D_X \frac{\partial^2 X}{\partial z^2} + k_{M1}AY - k_{M2}XY + k_{M3}AX - 2k_{M4}X^2
$$

(2)

$$
\frac{\partial Y}{\partial t} = D_Y \frac{\partial^2 Y}{\partial z^2} - k_{M1}AY - k_{M2}XY
$$

where $A = [BrO_3^-]$ = constant, $X = [HBrO_2] = X(z,t)$, $Y = [Br^-] = Y(z,t)$, z = space variable (one-dimensional reaction vessel, e.g. long, narrow tube), t = time, and D_X, D_Y are diffusion constants. Boundary conditions appropriate for a trigger wave propagating from right to left are

$$
X(-\infty,t) = 0 , \quad Y(-\infty,t) = [Br^-]_{ahead}
$$

(3)

$$
X(+\infty,t) = [HBrO_2]_{behind} , \quad Y(+\infty,t) = 0 .
$$

Field and Noyes estimate that $[Br^-]_{ahead} \approx 3 \times 10^{-5}M$. From Eqs. (2a) and

(3b) we estimate that

$$[HBrO_2]_{behind} = \frac{k_{M3}A}{2k_{M4}} \, .$$

As a first approximation to Eq. (2) with boundary conditions (3), consider only the diffusion, autocatalytic production, and disproportionation of $HBrO_2$:

$$\frac{\partial X}{\partial t} = D_X \frac{\partial^2 X}{\partial z^2} + k_{M3}AX - 2k_{M4}X^2$$

$$X(-\infty,t) = 0 \, , \, X(+\infty,t) = k_{M3}A/2k_{M4} \, .$$

Introducing the new variables

$$u = (2k_{M4}/k_{M3}A)X$$

$$\tau = (k_{M3}A) \, t$$

$$\zeta = \sqrt{k_{M3}A/D_X} \, z$$

(see Problem II.2), we cast our equation in dimensionless form (Fisher, 1937)

$$\frac{\partial u}{\partial \tau} = \frac{\partial^2 u}{\partial \zeta^2} + u(1 - u)$$

(4)

$$u(-\infty,\tau) = 0 \, , \, u(+\infty, \tau) = 1 \, .$$

To look for a solution of system (4) representing a wave of constant shape travelling from right to left with velocity $c > 0$, we substitute

$$u(\zeta,\tau) = \bar{u}(\phi) \, , \, \phi = \zeta + c\tau$$

into (4):

$$(5) \qquad c\frac{d\bar{u}}{d\phi} = \frac{d^2\bar{u}}{d\phi^2} + \bar{u}(1-\bar{u}) \;,\; \bar{u}(-\infty) = 0 \;,\; \bar{u}(+\infty) = 1 \;.$$

It is convenient to consider (5) in the phase plane (x,y), where $x = \bar{u}$, $y = d\bar{u}/d\phi$. Symbolizing differentiation with respect to phase ϕ by a prime, we write Eq. (5) as a pair of first order DE:

$$x' = y$$
$$(6)$$
$$y' = cy - x(1-x) \;.$$

We are looking for a ("heteroclinic") orbit connecting the critical points $(0,0)$ and $(1,0)$.

Linear stability analysis (see Chapter I) shows that $(1,0)$ is a saddle point and $(0,0)$ is an unstable node for

$$(7) \qquad c \geq c^* = 2 \;.$$

For $c < c^*$, $(0,0)$ is an unstable spiral point and, if $x \to 0$ as $\phi \to -\infty$, then $x < 0$ for some values of ϕ. Since x is a chemical concentration, which cannot be negative, Eq. (4) does not admit chemically realistic travelling wave solutions of constant shape for $c < 2$. For $c \geq c^*$, when $(0,0)$ is an unstable node, there do exist such travelling wave solutions. The existence proof is the subject of the following problem.

Problem 4. For $c \geq 2$ show that there exists one and only one solution of Eq. (6) for which

$$(x,y) \to (0,0) \quad \text{as} \quad \phi \to -\infty$$

$$(x,y) \to (1,0) \quad \text{as} \quad \phi \to +\infty \;.$$

Furthermore, show that

$$0 < x(\phi) < 1 \; , \; 0 < y(\phi) < \frac{c - \sqrt{c^2 - 4}}{2} \quad \text{for} \quad -\infty < \phi < +\infty \, .$$

Hint. Integrate (6) <u>backwards</u> in ϕ . Show that the phase plane has the appearance

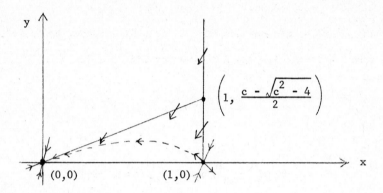

See Aronson and Weinberger (1975) Theorem 4.2.

By a statistical analysis of the growth and random walk processes at the leading edge of a wave (where u is very close to zero), Fisher (1937) argued that only waves with $c = c^* = 2$ would be observed. (Kolmogorov, Petrovsky and Piscounoff (1937) and Aronson and Weinberger (1975) reached the same conclusion on the basis of rigorous analysis of the partial differential equation.) Translating back to dimensional variables, we find that the trigger wave propagates at speed

$$c_d^* = \frac{dz}{dt} = \sqrt{k_{M3} A D_X} \; c^* = 2\sqrt{k_{R5} [H^+][BrO_3^-]D_X} \, .$$

Since $k_{R5} \approx 10^4 M^{-2} sec^{-1}$ and $D_X \approx 10^{-5} cm^2 sec^{-1}$,

$$c_d^* \approx 0.6 \; cm \, sec^{-1} M^{-1}\sqrt{[H^+][BrO_3^-]} \, .$$

Tilden (1974) has given a very different derivation of the same result. Field and Noyes (1974b) determined experimentally that

$$c = 0.04 \text{ cm}^2 \text{ sec}^{-1} M^{-1} \sqrt{[H^+][BrO_3^-]} \ .$$

From model (4) we get the correct functional dependence on $[BrO_3^-]$ and $[H^+]$, but the proportionality constant is too large.

Murray (1974b) has removed most of this discrepancy by including Br^- in the model. To the list u, ζ, τ , Murray adds the variable

$$v = (k_{M2}/k_{M3}Ar)Y$$

where

$$r = (k_{M2}/k_{M3}A)[Br^-]_{ahead} \approx 6/[BrO_3^-] \ .$$

In these terms, Eqs. (2) - (3) become

$$\frac{\partial u}{\partial \tau} = \frac{\partial^2 u}{\partial \zeta^2} + u(1 - u - rv)$$

(8)

$$\frac{\partial v}{\partial \tau} = \frac{\partial^2 v}{\partial \zeta^2} - buv$$

$$u(-\infty, \tau) = 0 \ , \ v(-\infty, \tau) = 1$$

(9)

$$u(+\infty, \tau) = 1 \ , \ v(+\infty, \tau) = 0$$

where

$$b = k_{M2}/2k_{M4} \approx 25[H^+] \ .$$

In going from Eq. (2) to Eq. (8) Murray neglects the term $k_{M1}AY$, which is small with respect to the other terms except at the leading edge of a wave, where $u = 0$, $v \approx 1$.

Murray proves the interesting theorem that the wave speed $c = \lambda(b,r)$, for all propagating wave front solutions of Eqs. (8) - (9), satisfies

$$c = \lambda(b,r) \leq 2$$

for all $b \geq 0$, $r \geq 0$. For $b \approx 5\text{-}15$ and $r \approx 20\text{-}50$, corresponding to the experimental conditions of Field and Noyes, Murray found numerically that $\lambda(b,r) \approx 0.2 - 0.4$. In dimensional terms

$$c_d \approx 0.08 - 0.16 \text{ cm sec}^{-1} M^{-1} \sqrt{[H^+][BrO_3^-]}$$

which differs from the experimental result by a factor of $2 - 4$.

Problem 5. Show that, for $b = 1 - r$, system (8) - (9) has a travelling wave solution of speed $c = \lambda(b, 1 - b) = 2\sqrt{b}$.
(Hint: let $v = 1 - u$.) Furthermore, show that

$$\lambda(0,r) = 0 \text{ , } r > 0 \text{ ; } \lambda(\infty,r) = 2 \text{ , } r \geq 0$$

$$\lambda(b,0) = 2 \text{ , } b > 0 \text{ ; } \lambda(b,\infty) = 0 \text{ , } b \geq 0 \text{ .}$$

See Murray (1974b).

Scroll waves.

Winfree (1973, 1974b, 1974c) has described several wave patterns: spirals elongated spirals, elongated rings and scroll rings; all of which he attributes to three dimensional scroll waves. These patterns have all nearly the same frequency $(1 \text{ min}^{-1} \pm 10\%)$. Waves propagate into virgin medium at about

6 mm/min but somewhat slower when following a previous wave. Elongated sources often break up into more symmetric sources. For example, an elongated ring source may break up into a pair of oppositely rotating spirals, or an elongated spiral may break up into a less-elongated ring pattern and a symmetric spiral. In all these transformations (except for the disappearance of a spiral on collision with an edge of the dish), "parity" is conserved, where parity = # clockwise rotating spirals - # counterclockwise rotating spirals, rings having parity = 0 . Elongated sources also tend to contract in length. Elongated rings, upon just reaching perfect symmetry, disappear.

That these waves all have the same period, Winfree takes as compelling evidence that they are all manifestations of the same thing: a three-dimensional scroll wave whose axis threads through the thin layer of medium from one inter- face to another. What we see is a projection of the scroll wave, which lies at various angles in the medium. On this basis the various transformations described in the last paragraph are explained as in Fig. 5.

It is not immediately obvious how scroll waves arise from reaction-diffusion equations (Chapter I)

$$(10) \qquad \frac{\partial x}{\partial t} = D \nabla^2 x + f(x) .$$

Winfree has (1974a) presented an interesting model of excitable kinetics, $f(x)$, and computed spiral wave solutions to Eq. (10) in two spatial dimensions. He borrows the kinetics from the theory of action potentials in nerve membranes (FitzHugh, 1961; McKean, 1970)

$$x = \begin{pmatrix} A \\ B \end{pmatrix}, \quad f(x) = \begin{pmatrix} -A - B + H(A - .05) \\ \frac{1}{2} A \end{pmatrix}$$

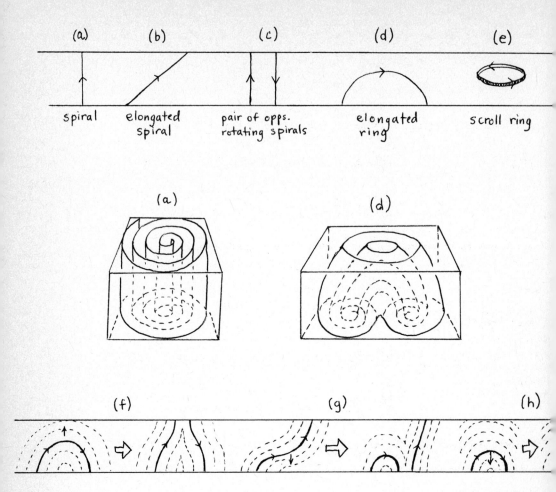

Fig. 5. Scroll waves in a thin layer of Z reagent (from Winfree, 1974b).
(a) When the scroll axis stands upright in the reagent, one sees a symmetric
spiral. In conjunction with the right-hand rule, the little arrow specifies
this as a counterclockwise rotating spiral. (b) When the scroll axis is tilte
one sees an elongated spiral. (c) A pair of oppositely oriented axes
corresponds to a pair of oppositely rotating spirals. (d) The vertical
projection of a scroll wave which bends from one interface back again is an
elongated ring. (e) If the scroll axis is bent into a ring lying horizon-
tally in the medium, one sees pairs of circular waves emitted from a ring-
like source, one propagating inward and one propagating outward. If we
imagine the scroll axis meandering through the reagent, we can understand
the parity conservation law. In (f) - (h) the dark line represents the scroll
axis and the dotted line represents emerging waves. In (f) an elongated
ring breaks up into a pair of oppositely rotating less-elongated spirals.
In (g) an elongated spiral breaks up into a less-elongated ring and a less-
elongated spiral. In (h) an elongated ring source becomes more and more
symmetric before completely disappearing.

where $H(x)$ is the step function

$$H(x) = \begin{cases} 1 , & x > 0 \\ 0 , & x \leq 0 \end{cases} .$$

Assuming a diagonal diffusion matrix as usual, Eqs. (10) become

(11)
$$\frac{\partial A}{\partial t} = D_A \nabla^2 A - A - B + H(A - .05)$$

$$\frac{\partial B}{\partial t} = D_B \nabla^2 B + \frac{1}{2} A$$

which are to be solved on the square $(x,y = \text{space variables})$

$$\mathbb{S} = \{(x,y) \,|\, 0 \leq x \leq L , \, 0 \leq y \leq L\}$$

with "no-flux" boundary conditions

$$\frac{\partial A}{\partial x} = \frac{\partial B}{\partial x} = 0 \quad \text{at} \quad x = 0 , \quad x = L$$

(12)

$$\frac{\partial A}{\partial y} = \frac{\partial B}{\partial y} = 0 \quad \text{at} \quad y = 0 , \quad y = L .$$

Winfree assumes that $D_A = D_B$. By sealing the spatial variables appropriately, we can set $D_A = D_B = 1$. Eqs. (11) in one spatial dimension (nerve axon) with $D_B = 0$ have been solved analytically by Rinzel and Keller (1973).

Without diffusion, Eqs. (11) reduce to a pair of piecewise linear ordinary differential equations

$$\frac{dA}{dt} = -A - B + H(A - \epsilon)$$

(13)

$$\frac{dB}{dt} = \frac{1}{2}A$$

Problem 6. For $\epsilon > 0$, show that the origin is a locally asymptotically stable focus.

Problem 7. For $\epsilon = 0$, show that there exists a unique orbitally asymptotically stable limit cycle. See Problem I.5.

Ans. (Andronov, Vitt and Khaikin, 1966, pp. 468-480)

$$\begin{pmatrix} A \\ B \end{pmatrix} = \frac{1 + e^{-\pi}}{1 - e^{-2\pi}} e^{-\frac{1}{2}t} \begin{pmatrix} -2 \sin \frac{1}{2}t \\ \cos \frac{1}{2}t + \sin \frac{1}{2}t \end{pmatrix} \qquad 0 < t < 2\pi$$

$$\begin{pmatrix} A \\ B - 1 \end{pmatrix} = \frac{1 + e^{-\pi}}{1 - e^{-2\pi}} e^{-\frac{1}{2}(t - 2\pi)} \begin{pmatrix} 2 \sin \frac{1}{2}(t - 2\pi) \\ -\cos \frac{1}{2}(t - 2\pi) - \sin \frac{1}{2}(t - 2\pi) \end{pmatrix} \qquad 2\pi < t < 4\pi$$

Problem 8. For ϵ sufficiently small, show that there exists a stable limit cycle close to the one found for $\epsilon = 0$. There must be an unstable limit between the origin and the stable limit cycle. Find it.

The phase plane for system (13) with $\epsilon = .05$ is illustrated in Fig. 6. It bears some similarity with the excitable Oregonator illustrated in Fig. 3, if we identify $A \sim Ce^{+4}$ and $B \sim Br^-$.

Without reaction, we have just a pair of heat equations

(14) $$\frac{\partial A}{\partial t} = \nabla^2 A \ , \qquad \frac{\partial B}{\partial t} = \nabla^2 B \ .$$

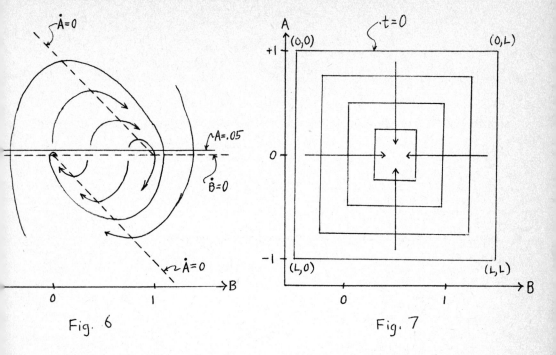

Fig. 6. Phase plane for Eq. (13) with $\epsilon = 0.05$. The origin is globally stable, but perturbations which drive $A > .05$ result in a long excursion before returning to the vicinity of the steady state.

Fig. 7. Solution of Eq. (14) in A,B plane subject to boundary conditions (12) and initial conditions (15). At $t = 0$ the x,y plane is mapped into the A,B plane as indicated in the figure. As t increases, $A \to 0$ and $B \to \frac{1}{2}$; the image of the x,y plane shrinks like an elastic membrane.

With the no-flux boundary conditions (12) and initial conditions of crossed

concentration gradients

$$(15) \qquad A(x,y,0) = A(x) = 1 - 2x/L$$

$$B(x,y,0) = B(y) = -\frac{1}{2} + 2y/L$$

the solution of (14) looks like a shrinking elastic membrane in the A,B

plane, Fig. 7. (See Problem I.10.)

Combining reaction and diffusion the image of the reaction medium in the phase plane $(x,y \to A,B)$ behaves like an elastic membrane embedded in a viscous medium flowing according to the local kinetics (Winfree, 1974a; for this analogy to be valid it is essential that $D_A = D_B$). Diffusion makes the membrane contract but reaction keeps flinging it out from the origin. After a transient period, Winfree's numerical calculations seem to settle on a rotating spiral wave illustrated in Fig. 8.

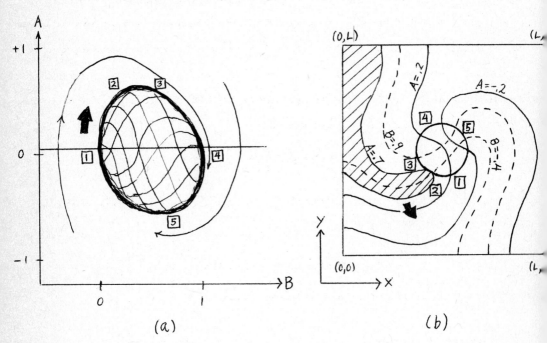

(a)   (b)

Fig. 8. Snapshot of a rotating solution of Eq. (11) computed by Winfree (1974a) (a) Phase plane. (b) Real space (L = 50). As t increases, the pattern in (b) simply rotates counterclockwise. Within the "core", bounded by the heavily outlined circle in (b), concentration isobars of A and B are criss-crossed. In the phase plane, the core is stretched across the center of a sequence of states [1] - [5] . All volume elements outside the core traverse this sequence periodically. Volume elements close to the steady state [1] are pulled across threshold by their neighbors [1.5] and become excited [2] . Arbitrarily callir

$A > .7$ BLUE and $A < .2$ RED , we find the leading edge of a blue band at $\boxed{2}$ propagating into a red medium. After excitation, the medium is refractory for a time $\boxed{3}$ - $\boxed{5}$ and then recovers to the vicinity of the steady state, only to be excited once again.

The rotating pattern in Fig. 8(b) looks similar to the tip of a spiral wave in Z reagent. Surprisingly enough the size of the pattern is of the correct order of magnitude. For in this model the period of rotation is about $4\pi \approx 12.5$ time units. Using 15 sec as the rotation period in Z reagent, we have 1 time unit $\approx$ 1 sec . Furthermore, $D = 1$ (space unit)^2 (time unit)$^{-1} \approx 10^{-5}$ cm^2 sec^{-1} , which implies that 1 space unit $\approx 3 \times 10^{-3}$ cm . Now, from Fig. 8(b), since $L = 50$, the diameter of the core is approximately 13 space units $\approx .4$ mm . It is observed that the tip of a spiral wave wanders around in a spatial domain approximately $\frac{1}{2}$ mm in diameter. Alternatively, this could be expressed as the wave velocity

$$v = \frac{\pi \cdot \text{diameter}}{\text{period}} \approx \frac{5/4 \text{ mm}}{1/4 \text{ min}} = 5 \frac{\text{mm}}{\text{min}} \quad,$$

which is approximately the observed value. During one rotation, a molecule the size of Br$^-$ diffuses a distance

$$d = \sqrt{2DT} \approx .2 \text{ mm} \approx \text{core radius} \quad,$$

so diffusion cannot be neglected within the core.

Plane wave and spiral wave solutions of reaction-diffusion equations.

Several more rigorous treatments of wave-like solutions to reaction-diffusion equations

$$(10) \qquad \frac{\partial x}{\partial t} = D \nabla^2 x + f(x)$$

have recently appeared.

Far from a source, chemical waves look like plane waves, i.e.

$$(16) \qquad x(\vec{r}, t) = y(\omega t - \vec{k} \cdot \vec{r})$$

where x = n-vector of chemical concentrations , $\vec{r}$ = vector of spatial coordinates , t = time , ω = angular frequency > 0 , $\vec{k}$ = wave number vector , and y = 2π-periodic function of its argument , $\omega t - \vec{k} \cdot \vec{r} = \psi$ = phase .

Kopell and Howard (1973b) have looked for solutions of this form to reaction-diffusion equations, such as (10). Plugging (16) into (10) we see that $y(\psi)$ must satisfy the system of second order ordinary differential equations

(17)
$$\omega \frac{dy}{d\psi} = k^2 D \frac{d^2 y}{d\psi^2} + f(y) .$$

Let f(0) = 0 , that is, shift the steady state to the origin.

Let us state a few theorems due to Kopell and Howard. First we restrict ourselves to two-component systems $x = (x_1, x_2)$. As usual, I = identity matrix .

Theorem 1. Let the two-dimensional vector field f(x) have an unstable focus at x = 0 , that is, the Jacobian matrix, $M = f_x(0)$, has eigenvalues $p \pm iq$ ($p > 0$, $q > 0$) . Let $M = p(I + M_1)$, so that $\text{Tr} M_1 = 0$. Let the 2×2 positive definite matrix D be written as $D = \frac{1}{2}(\text{Tr } D)(I + \delta D_1)$, where D_1 is a real symmetric matrix with eigenvalues ± 1 and $\delta \geq 0$. If

(18)
$$2\delta < \text{Tr}(M_1 D_1) + \sqrt{(\text{Tr}(M_1 D_1))^2 + 4(q/p)^2} ,$$

then there exists a one-parameter family of plane wave solutions of Eq. (10).

To prove this theorem, Kopell and Howard use the Hopf bifurcation theorem (Hopf, 1942) to prove the existence of a one-parameter family of 2π-periodic solutions of the fourth order system (17). The bifurcation parameter they use is k^2/ω . The solutions in the one-parameter family (parametrized by amplitude) all have wave numbers near k_0 , where k_0 is the value of k for which $M - k^2 D$ has conjugate pure imaginary (non-zero) eigenvalues. k_0

exists and is unique, if and only if inequality (18) is satisfied.

The theorem can be extended to n-component systems, $n > 2$, for the diffusion matrix D in some open neighborhood of the scalar matrices dI , $0 < d < \infty$. For $n = 2$ this neighborhood is precisely described by inequality (18), but as yet there is no similar characterization for $n > 2$.

If a chemical system has an unstable focus, as hypothesized in Theorem 1, it is natural to expect periodic solutions as well. That case is covered by

Theorem 2. Suppose the n-dimensional system $x' = f(x)$ has a stable non-constant periodic solution $x = \tilde{x}(t)$, with period $2\pi/\omega_0$. Then, for sufficiently small k^2 , there exists a one-parameter family of plane wave solutions of Eq. (10) with uniquely defined frequency $\omega = \Omega(k^2)$ close to ω_0 .

To prove this theorem (actually they prove a somewhat stronger result), Kopell and Howard use an iterative procedure to prove the existence of 2π-periodic solutions of the 2n-dimensional system (17), which are close to the known 2π-periodic solution of $\omega_0 y' = f(y)$, to which (17) reduces as $k^2 \to 0$. The proof suggests a convenient procedure for calculating plane wave solutions, which Kopell and Howard tested on the Brusselator (see Chapter I, pp. 15ff). Fig. 9a displays the amplitude and the dispersion relation $\omega = \Omega(k^2)$ calculated for the Brusselator under the conditions $a = 1$, $b = 2.8$ and 3.6 , $D = I$. Fig. 9b displays the dispersion relation measured by Tatterson and Hudson (1973) for waves propagating along a tube of Belousov-Zhabotinskii reagent.

Kopell and Howard also investigated the stability properties of these plane waves as solutions of the partial differential equations (10). The plane waves discussed in Theorem 1, coming from a Hopf bifurcation, have small amplitude close to the point of bifurcation, in which case applies

Theorem 3. If D is sufficiently close to I and the plane wave solution

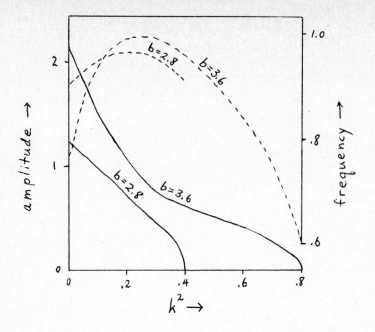

Fig. 9a. Amplitude (——) and frequency (----) as functions of wave number
(k = 2π/λ , λ = wavelength) for plane wave solutions of Eq. (3) with
Brusselator kinetics. Computed from Table I of Kopell and Howard (1973b),
with amplitude defined as y - b for y given in the table. The other
parameters are fixed as a = D_x = D_y = 1 .

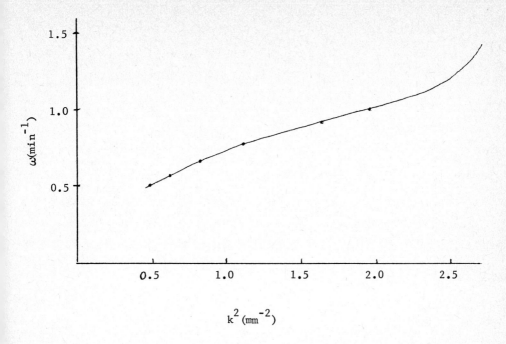

Fig. 9b. Dispersion relation for Belousov-Zhabotinskii reagent (after
          Fig. 6 in Tatterson and Hudson, 1973).

$x(r,t) = y(\omega t - kr)$ has sufficiently small amplitude, then the solution is
unstable.

The plane waves discussed in Theorem 2 have amplitudes close to the
amplitude of the limit cycle, $\tilde{x}(t)$, as $k^2 \rightarrow 0$. Kopell and Howard conjecture
that, if $x' = f(x)$ has a stable limit cycle and D is sufficiently close
to I , then there is a range of stable plane wave solutions for k small.

Kopell and Howard have written several nice expository papers on
travelling wave solutions of reaction-diffusion equations (e.g. Kopell
and Howard, 1974, 1975). One result which is particularly worth mentioning
is that one can construct axisymmetric periodic travelling waves which are
completely regular at the origin (see also Greenberg, 1975). That is,

mathematically speaking there is no necessity for a speck of dust at the center of a target pattern, though Winfree's experiments argue strongly for its presence.

Ortoleva and Ross (1974) have also presented an overview of chemical wave phenomena.

Do there exist spiral wave solutions of Eq. (10)? Consider the reaction-diffusion equation in a polar coordinate system: $x = x(r,\theta,t) \in R^n$. Let $D = \text{diag}(D_1,\ldots,D_n)$ and furthermore assume that $D_1 = \ldots = D_n$, that is, D is just a constant times the identity matrix. Assume that the reaction equations $x' = f(x)$ have a limit cycle solution of period $T = 2\pi/\omega_0$. Introduce dimensionless time and space variables

$$\tau = \omega_0 t , \qquad \rho = kr$$

where k (a reciprocal length) is to be fixed later. Then Eq. (10) becomes

$$\omega_0 \frac{\partial x}{\partial \tau} = k^2 D \nabla^2 x + f(x) , \quad \text{or}$$

(10′)
$$\frac{\partial x}{\partial \tau} = \kappa^2 \nabla^2 x + g(x)$$

where $\kappa^2 = \dfrac{k^2 D}{\omega_0} > 0$, $g(x) = \dfrac{1}{\omega_0} f(x)$. Eq. (10′) is now completely dimensionle

The one-dimensional version of Eq. (10′) is

(19)
$$\frac{\partial x}{\partial \tau} = \kappa^2 \frac{\partial^2 x}{\partial \xi^2} + g(x) .$$

Since $dx/d\tau = g(x)$ has a 2π-periodic solution, we know from Theorem 2 that Eq. (19) has a one-parameter family of plane wave solutions

$$x(\xi,\tau) = y(\xi + \omega\tau ; \kappa^2).$$

for κ^2 sufficiently small and a unique $\omega = \Omega(\kappa^2) = 1 + \mathbb{O}(\kappa^2)$. In particula $y(\psi ; k^2)$ is a 2π-periodic function of ψ which satisfies

(17′)
$$\Omega(\kappa^2)\,\frac{dy}{d\psi} = \kappa^2\,\frac{d^2y}{d\psi^2} + g(y) \ .$$

Moreover,

$$y(\psi;\,k^2) = \tilde{x}(\psi) + \Theta(\kappa^2) \ ,$$

where $\tilde{x}(\tau)$ is the 2π-periodic solution of $dx/d\tau = g(x)$.

How do we characterize a spiral wave? Of most importance, time translation must be equivalent to rotation, that is,

$$x(\rho,\,\theta,\,\tau) = u(\rho,\,\phi) \ , \qquad \phi = \theta + \omega\tau \ .$$

The function $u(\rho,\phi)$ satisfies

(20)
$$\omega\,\frac{\partial u}{\partial \phi} = \kappa^2\,\nabla^2 u + g(u) \ , \qquad \nabla^2 = \frac{\partial^2}{\partial\rho^2} + \frac{1}{\rho}\frac{\partial}{\partial\rho} + \frac{1}{\rho^2}\frac{\partial^2}{\partial\phi^2} \ .$$

Furthermore, we look for a function $\Phi(\rho)$, representing the asymptotic level lines

$$\lim_{\rho\to\infty} u(\rho,\Phi(\rho)) = \text{constant} \ ,$$

with the properties that

$$\lim_{\rho\to\infty}\frac{d\Phi}{d\rho} = 1 \ , \qquad \lim_{\rho\to\infty}\frac{d^2\Phi}{d\rho^2} = 0 \ .$$

That is, asymptotically the pitch of the spiral is unity. (See DeSimone, Beil and Scriven, 1973, for a discussion of the geometry of spiral waves in the context of reaction-diffusion equations.) In real space, $r = \rho/k$, the wavelength of the spiral pattern tends asymptotically to $2\pi/k$, which fixes the length scale introduced earlier. Finally, as we go around a large circle from $(\rho,\,\Phi(\rho))$ back again, we should see a 2π-periodic function; that is,

$$\lim_{\rho\to\infty} u(\rho,\,\Phi(o) + \alpha) = u_\infty(\alpha) \ ,$$

where $u_\infty(\alpha)$ is a 2π-periodic function of α. To describe a spiral wave solution we must specify:

(i)   the solution of Eq. (20), $u(\rho,\phi)$.

(ii)   the frequency, ω.

(iii) the asymptotic level lines, $\Phi(\rho)$.

(iv)   the limit function, $u_\infty(\alpha)$.

For spiral waves observed in Z reagent, we have already remarked that the wavelength $\sim 1.5\,mm$ and the period $\sim 15\,sec$. Thus $k \sim 50\,cm^{-1}$, $\omega_0 \sim .5\,sec^{-1}$, and, since $D \sim 10^{-5}\,cm^2\,sec^{-1}$,

$$\kappa^2 \sim .05 .$$

This suggests that it would be reasonable to look for an asymptotic representation of spiral wave solutions in powers of κ^2. Greenberg (1975) has obtained the following results.

Theorem 4. If $\Phi_0(\rho)$ satisfies $\lim\limits_{\rho\to\infty} d\Phi_0/d\rho = 1$ and

(21)
$$\frac{1}{\rho}\frac{d}{d\rho}\left(\rho\,\frac{d\Phi_0}{d\rho}\right) = \Omega\left(\kappa^2\left[\frac{1}{\rho^2} + \left(\frac{d\Phi_0}{d\rho}\right)^2\right]\right) - \Omega(\kappa^2) ,$$

then, for sufficiently small κ^2, there exist solutions of Eq. (10′) which are asymptotically spiral waves described by

(i) $u(\rho,\phi) = y\left(\phi - \Phi_0(\rho)\,;\, \kappa^2\left[\frac{1}{\rho^2} + \left(\frac{d\Phi_0}{d\rho}\right)^2\right]\right) + \Theta(\kappa^2)$

(ii) $\omega = \Omega(\kappa^2) = 1 + \Theta(\kappa^2)$

(iii) $\Phi(\rho) = \Phi_0(\rho) + \Theta(\kappa^2)$

(iv) $u_\infty(\alpha) = y(\alpha;\kappa^2) = \tilde{x}(\alpha) + \Theta(\kappa^2)$.

To prove this theorem Greenberg introduces the function

$$u^{\#}(r,\alpha) \equiv u(\rho\,,\,\Phi(\rho) + \alpha)$$

which is 2π-periodic in α and satisfies

$$\left[\omega + \frac{\kappa^2}{\rho}\frac{d}{d\rho}\left(\rho\frac{d\Phi}{d\rho}\right)\right]\frac{\partial u^{\#}}{\partial\alpha} = \kappa^2\left[\frac{1}{\rho^2} + \left(\frac{d\Phi}{d\rho}\right)^2\right]\frac{\partial^2 u^{\#}}{\partial\alpha^2} + g(u^{\#})$$

(22)

$$+ \kappa^2\left(\frac{\partial^2}{\partial\rho^2} - 2\frac{d\Phi}{d\rho}\frac{\partial^2}{\partial\alpha\partial\rho} + \frac{1}{\rho}\frac{\partial}{\partial\rho}\right)u^{\#} \quad .$$

Furthermore,

$$\lim_{\rho\to\infty} u^{\#}(\rho,\alpha) = u_{\infty}(\alpha) \quad .$$

If we insist for mathematical reasons that

$$\lim_{\rho\to\infty}\frac{\partial^{m+n}u^{\#}}{\partial\rho^m\partial\alpha^n} = 0 \; , \quad m \geq 1 \; , \quad n \geq 0 \; ,$$

then, in the limit as $\rho \to \infty$, Eq. (22) becomes

$$\omega\frac{du_{\infty}}{d\alpha} = \kappa^2\frac{d^2 u_{\infty}}{d\alpha^2} + g(u_{\infty}) \quad ,$$

which is just the plane wave equation (17′). Thus we choose

$$\omega = \Omega(\kappa^2) \quad \text{and} \quad u_{\infty}(\alpha) = y(\alpha\,;\,\kappa^2) \quad .$$

Now assume that $\Phi(\rho)$ and $u^{\#}(\rho,\alpha)$ admit the asymptotic expansions

$$\Phi(\rho) = \Phi_0(\rho) + \kappa^2\Phi_1(\rho) + \dots$$

$$u^{\#}(\rho,\alpha) = u_0(\rho,\alpha) + \kappa^2 u_1(\rho,\alpha) + \dots$$

Substituting these expressions into Eq. (22) and keeping only the leading order terms, we obtain

$$\left[\Omega(\kappa^2) + \frac{\kappa^2}{\rho}\frac{d}{d\rho}\left(\rho\frac{d\Phi_0}{d\rho}\right)\right]\frac{\partial u_0}{\partial\alpha} = \kappa^2\left[\frac{1}{\rho^2} + \left(\frac{d\Phi_0}{d\rho}\right)^2\right]\frac{\partial^2 u_0}{\partial\alpha^2} + g(u_0) \quad ,$$

which is again just the plane wave equation (17′). Thus,

$$u_0(r,\alpha) = y\left(\alpha; \kappa^2\left[\frac{1}{\rho^2} + \left(\frac{d\Phi_0}{d\rho}\right)^2\right]\right)$$

is an appropriate solution, provided $\Phi_0(\rho)$ satisfies Eq. (21). Obviously the dispersion relation $\omega = \Omega(\kappa^2)$ for plane waves plays a prominent role in the theory of spiral wave solutions. Recall Fig. 9, which records $\Omega(k^2)$ for the Brusselator.

It should be remarked that the asymptotic expansions for spirals waves developed here are not regular at the origin. We saw the same problem in Chapter I, see Eq. (I.22). At the moment, the core of a spiral wave still eludes analysis. Another misfortune of this theory is the existence of a continuous family of spiral waves for all wavenumbers k sufficiently small. Only spiral waves of fixed wavenumber (pitch) are observed. An adequate explanation of this fact has also yet to appear.

Under certain reasonable conditions on the dispersion relation, Greenberg (private communication) has shown that

$$\rho\,\frac{d\Phi_0}{d\rho} \to \rho + \sigma \quad \text{as} \quad \rho \to \infty \; ;$$

where σ is a constant which can be calculated given the function $\Omega(\kappa^2)$. Thus

$$\Phi_0(\rho) \to \rho + \sigma\ell n\rho \quad \text{as} \quad \rho \to \infty$$

and in general the spiral level lines have a logarithmic as well as "Archimedean" part.

We might paraphrase Theorems 2 and 4 loosely as: limit cycles $\Rightarrow$ plane waves $\Rightarrow$ spiral waves. Furthermore, we might extend the conjecture of Kopell and Howard as follows: if $x' = f(x)$ has a stable limit cycle and D is sufficiently close to I, then for κ sufficiently small there exist stable rotating solutions of Eq. (10$'$) which look like Archimedean spirals far from the origin.

Recalling the phenomenon of hard self-excitation in the Oregonator (Chapter III), we have every reason to suspect that, besides a stable homogeneous resting state, there exist stable rotating spiral waves. Furthermore, knowing the $2\pi/\omega_0$-periodic solution $\tilde{x}(t)$ of the reaction equations as well as we do, we can immediately describe the spiral wave solution far from the origin:

$$(23) \qquad x(r,\theta,t) \approx \tilde{x}(\theta + \omega_0 t - kr - \sigma \ln kr) \qquad \text{for } r \text{ large .}$$

From Fig. III. 4 we construct Figs. 10 and 11, which illustrates approximately the concentrations of $HBrO_2$, Br^- and Ce^{+4} in a spiral wave.

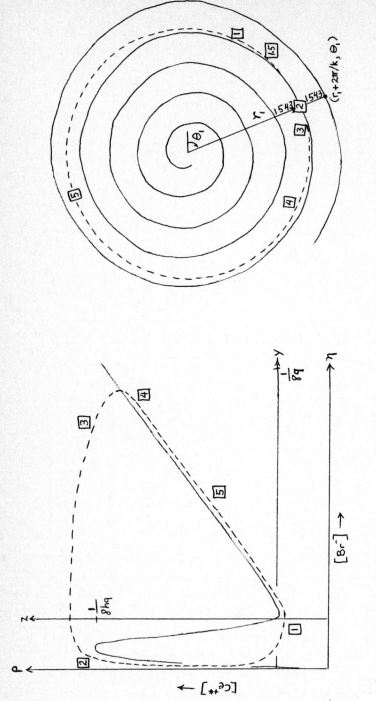

Fig. 10. Snapshot ($t = t_1$) of spiral wave solution of Oregonator equations predicted by Eq. (23) and Fig. 8. Compare with Fig. 8. and Fig. III.4. At fixed radius ($r = r_1 \gg 1/k$), $x(r_1, \theta, t_1) = \tilde{x}(\theta + \text{constant})$ that is, the dashed circle in real space maps onto the limit cycle in phase space. At fixed angle ($\theta = \theta_1$), $x(r, \theta_1, t_1) = \tilde{x}(\text{constant} - kr)$ for $kr \gg 1$: the dashed line between (r_1, θ_1) and ($r_1 + 2\pi/k, \theta_1$) maps onto the limit cycle as well. Points [1] → [5] in real space map onto points [1] - [5] in the phase plane.

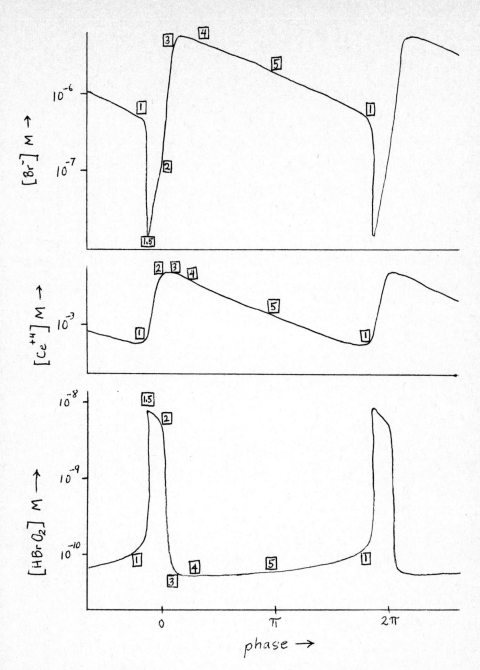

Fig. 11. Waveforms for Br^-, Ce^{+4} and $HBrO_2$ predicted by Eq. (23). Phase $= \theta + \omega_0 t - kr$, for r large. The limit cycle for the Oregonator equations was calculated for $\epsilon = .0002$, $q = .006$, $h = .75$, $p = 5$. Concentrations were prescribed by: $[Br^-] = (3.33 \times 10^{-7} M)\eta$, $[Ce^{+4}] = (1.67 \times 10^{-4} M)\rho$, $[HBrO_2] = (5 \times 10^{-11} M)\xi$.

Appendix. The Zhabotinskii-Zaikin-Korzukhin-Kreitser model.

The model

Before the appearance of the Field-Körös-Noyes mechanism, Zhabotinskii
and coworkers (1971) suggested a simple semi-empirical model of the BZ
reaction. For comparison with the Field-Noyes model discussed in
Chapter III, I shall write the Zhabotinskii-Zaikin-Korzukhin-Kreitser
(ZZKK) model as a set of hypothetical reactions. To be consistent with
the nomenclature used by the Russian authors, we let $A = BrO_3^-$,
$B = BrCH(COOH)_2$, $X = Ce^{+4}$, $Y = HBrO_2$ ("active intermediate"), $Z = Br^-$.
(Unfortunately this is a permutation of the definitions of X,Y,Z used
by Field and Noyes.) Furthermore, let $E = Br_2C(COOH)_2$ and/or Br_3CCOOH,
$P = HOBr$, $X' = Ce^{+3}$. Then the two models can be written as

ZZKK model		FN model[*]	
(L1)	$A + Y + X' \rightarrow 2Y + X$	(M3)	$A + Y \rightarrow 2Y + 2X$
(L2)	$Y + Z \rightarrow P + Z$	(M2)	$Y + Z \rightarrow 2P$
(L3)	$B + X \rightarrow X' + Z$	(M5)	$B + X \rightarrow hZ$
(L4)	$Z \rightarrow out$	(M1)	$A + Z \rightarrow Y + P$
(L5)	$A \rightarrow Y$		
(L6)	$E + X \rightarrow X + Z$		

[*]Remember, $Y = HBrO_2$, $Z = Br^-$, $X = Ce^{+4}$

We notice the following differences:

1. ZZKK assume that the rate of autocatalytic production of "active
intermediate" (presumably $HBrO_2$) is proportional to the concentration
of Ce^{+3}

$$r_{L1} = k_1 AY(C - X)$$

where $C = X + X'$ = total cerium concentration. (As usual I am using the symbols $X, Y, Z, \ldots$ to represent the concentrations of the chemical species as well as the names of the chemicals themselves.)

2. ZZKK assume that Br^- is not annihilated by $HBrO_2$. Some $HOBr$ formed in step (M2) probably oxidizes formic acid, releasing bromide (see Smith, 1972)

$$HOBr + HCOOH \rightarrow H^+ + Br^- + CO_2 + H_2O \ ,$$

but this reaction is sluggish compared with (R2), see p. 33.

3. FN lump together all uncertainty about Ce^{+4} oxidation of organic species into the stoichiometric parameter h. ZZKN separate Ce^{+4} oxidation of bromomalonic acid (L3) from Ce^{+4} induced decomposition of dibromomalonic and tribromomalonic acid (L6) - see Vavilin and Zhabotinskii (1969). For step (L3)

$$r_{L3} = k_3 BX \ ,$$

where $k_3 \approx 0.1 \ M^{-1} sec^{-1}$ for $B < 0.2M$ (Kasperek and Bruice, 1971). For step (L6) ZZKK assume a rate law

$$r_{L6} = k_6 A [k_7 Y - \alpha(A,B)]^2 X$$

where $k_6 \equiv 0.1 \ M^{-1} sec^{-1}$, $k_7 Y - \alpha$ is a dimensionless quantity and the function $\alpha(A,B)$ remains for the moment unspecified. This rate term was selected empirically in order to fit the observed waveform.

4. ZZKN assume separate reactions for a Br^- sink (L4) and an $HBrO_2$ source (L5).

From experimental rate studies ZZKK propose that step (L2) proceeds fastest, steps (L1) and (L4) are fast in comparison with (L3) and (L6), and step (L5) is the slowest. In terms of rate parameters, all with units $M^{-1}sec^{-1}$,

(1)
$$k_2 \approx k_4 k_7 \gg k_1 A \gg \frac{k_1}{k_7} \approx k_3 \approx k_6 \approx k_5 k_7 .$$

Notice that $k_7 A \gg 1$, i.e. k_7^{-1}, which is a characteristic concentration of active intermediate, is small with respect to $A \approx B \approx C$.

Regarding A,B,C as constants over the time of many oscillations, we write down three kinetic equations to describe model (L1) - (L6) :

(2)
$$\frac{dX}{dt} = k_1 AY(C - X) - k_3 BX$$

$$\frac{dY}{dt} = k_1 AY(C - X) - k_2 YZ + k_5 A$$

$$\frac{dZ}{dt} = k_3 BX - k_4 Z + k_6 A(k_7 Y - \alpha)^2 X .$$

Introducing the dimensionless variables (Othmer, 1975)

$$x = \frac{X}{C} , \quad y = k_7 Y , \quad z = \frac{k_4 Z}{k_5 C} , \quad \tau = \frac{k_1 A}{k_7} t$$

into Eq. (2), we obtain

(3)
$$\frac{dx}{dt} = y(1 - x) - \delta x$$

$$\epsilon \frac{dy}{dt} = y(1 - x) - \lambda_1 yz + \lambda_2$$

$$\mu \frac{dz}{dt} = \delta x + \lambda_3 (y - \alpha)^2 x - \lambda_4 z$$

where

$$\delta = \frac{k_3 k_7 B}{k_1 A} \;,\quad \epsilon = \frac{1}{k_7 C} \;,\quad \mu = \frac{k_5}{k_4} \;,$$

$$\lambda_1 = \frac{k_2 k_5}{k_1 k_4 A} \;,\quad \lambda_2 = \frac{k_5 k_7}{k_1 C} \;,\quad \lambda_3 = \frac{k_6 k_7}{k_1} \;,\quad \lambda_4 = \frac{k_5 k_7}{k_1 A}$$

are all dimensionless parameters. From Eq. (1) we obtain (for $A \approx C$)

$$\mu \ll \lambda_2 \approx \epsilon \ll \frac{\lambda_1 \lambda_3}{\lambda_4} \approx 1$$

and

$$\frac{\lambda_1 \delta}{\lambda_4} \approx \frac{B}{A} \;.$$

Consider the two-dimensional manifold defined by

$$z = \phi(x,y) \equiv \frac{\delta}{\lambda_4} x + \frac{\lambda_3}{\lambda_4} (y - \alpha)^2 x \;.$$

If $z > \phi(x,y)$, then $\mu dz/d\tau < 0$ and, since $\mu \ll \epsilon \ll 1$, $z(\tau)$ decreases very rapidly compared to $y(\tau)$ or $x(\tau)$. On the other hand, if $z < \phi(x,y)$, then $\mu dz/d\tau > 0$ and $z(\tau)$ increases rapidly. That is, after an initial relaxation time

$$\tau \sim \frac{\mu}{\lambda_4} = \frac{k_1 A}{k_4 k_7} \ll 1 \;,$$

we have

$$z(\tau) \approx \phi(x(\tau), y(\tau)) \;.$$

In this manner, Eq. (3) reduces to the planar system

$$\frac{dx}{d\tau} = P(x,y;\delta) \equiv y(1-x) - \delta x$$

(4)

$$\epsilon \frac{dy}{d\tau} = Q(x,y;\alpha,\epsilon) \equiv y\{1 - x[1 + \alpha + (y - \alpha)^2]\} + \epsilon$$

where we have assumed that $\alpha(A,B) = B/A$.

Although $\epsilon \ll 1$, Eq. (4) does not reduce further because, for certain values of α, the one dimensional manifold defined by $Q(x,y;\alpha,\epsilon) = 0$ is not stable. Under certain conditions on α and δ (with ϵ sufficiently small) one finds periodic solutions of Eq. (4) with fast "jumps". As in Chapter III, we shall study these relaxation oscillations by phase plane methods (Tyson, 1976).

Steady states

The point (x_0,y_0) in the phase plane is a steady state solution of DE(4) if and only if $P(x_0,y_0) = 0$ and $Q(x_0,y_0) = 0$. The first of these conditions defines an hyperbolic relation between x and y

(5a)
$$P(x,y) = 0 \Leftrightarrow x = \frac{y}{\delta + y},$$

whereas the second defines a more complicated relation

(5b)
$$Q(x,y) = 0 \Leftrightarrow x[1 + \alpha + (y - \alpha)^2] = 1 + \frac{\epsilon}{y}.$$

Substitution of (5a) into (5b) gives a fourth order equation in y

(6)
$$y^4 - 2\alpha y^3 + \alpha(1 + \alpha)y^2 - (\epsilon + \delta)y - \epsilon\delta = 0$$

For $0 < \epsilon \ll 1$, Eq. (6) has a real negative root

$$y = -\epsilon + \mathfrak{G}(\epsilon^2).$$

Factoring out this root and ignoring terms $\mathfrak{G}(\epsilon)$, we obtain from Eq. (6)

the cubic equation

(7)
$$y^3 - 2\alpha y^2 + \alpha(1 + \alpha)y - \delta = 0 .$$

Notice that Eq. (7) has no real negative roots because the left hand side is less than zero (indeed, less than $-\delta$) for all $y < 0$.

Let

$$\emptyset = \frac{1}{4} \delta^2 - \frac{1}{27} \alpha^2 (\alpha + 9)\delta + \frac{1}{27} \alpha^3 (\alpha + 1)^2 .$$

Then, if $\emptyset > 0$, Eq. (7) has one real positive root and a pair of complex conjugate roots. On the other hand, if $\emptyset < 0$, Eq. (7) has three real positive roots: that is, DE(4) admits more than one steady state solution in the positive quadrant. Multiple steady state (MSS) solutions will appear for all δ satisfying

$$\delta_- < \delta < \delta_+ ,$$

where

(8)
$$\frac{27}{2} \delta_\pm = \alpha^2 (\alpha + 9) \pm [\alpha(\alpha - 3)]^{3/2} .$$

For $\alpha < 3$, $\emptyset > 0$ for all real δ and Eq. (7) has only one real positive root. For $\alpha = 3$, $\delta = 8$, Eq. (7),reduces to $(y - 2)^3 = 0$. For $\alpha > 3$ there is a range of real values of δ for which Eq. (7) has three real positive roots. Plotted in the (α, δ) plane, the region of MSS behavior is cusp-shaped near $\alpha = 3$, $\delta = 8$. Notice that

(9)
$$\delta_- \to \alpha^2 \quad \text{and} \quad \delta_+ \to \frac{4}{27} \alpha^3 \quad \text{as} \quad \alpha \to \infty .$$

When plotted in the (log α, log δ) plane, as in Fig. 1, these expressions give excellent approximations to $\delta_{\pm}$.

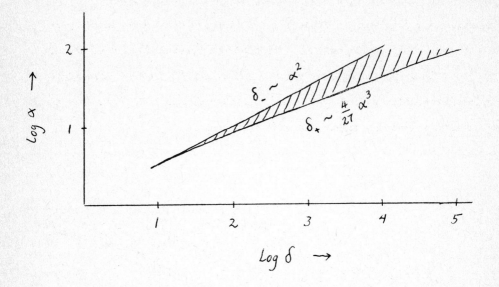

Fig. 1. Region of multiple steady state behavior in parameter space.

From the derivation of Eq. (4) we have $\alpha \approx \delta \approx B/A$. Obviously, for $\alpha = \delta$ one will not observe multiple steady states. However, in order to fit the waveform of periodic solutions of Eq. (4) to observed oscillations (see next section), ZZKK postulated that

(10)
$$\alpha = \frac{5}{4}(M^{-\frac{1}{2}})\sqrt{B}\,\frac{A + 0.1M}{A}\quad,\quad \delta = \frac{B}{2A}\quad,\quad \epsilon = \frac{5 \times 10^{-6}M}{C}$$

where M simply means "molar" and, as before, A = [BrO_3^-] , B = [BrMA] , C = [Ce^{+3}] + [Ce^{+4}] . For what values of A,B,C , if any, does (α,δ) given by Eq. (10) lie in the wedge in Fig. 1? From Eq. (10) we derive that

(11)
$$2 \log \alpha = \log \delta + \log \frac{50}{16} \frac{(A + .1)^2}{A} \ .$$

That is, for fixed A , the plot of $\log \alpha$ vs $\log \delta$ has slope $= \frac{1}{2}$ and intercept $= \frac{1}{2} \log (50(A + .1)^2/16A)$. Notice that the intercept $\rightarrow \infty$ for $A \rightarrow 0$ and $A \rightarrow \infty$. The smallest value of the intercept (~ 0.05) occurs at $A = 0.1M$. These observations are summarized in Fig. 2. Assuming a maximum bromomalonic acid concentration of approximately $2M$, we have cut off the region at the right hand side according to $\delta_{max} \approx 1/A$.

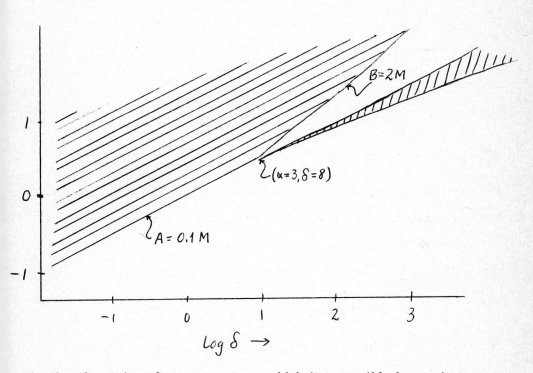

Fig. 2. The region of parameter space, which is accessible by varying
$A = [BrO_3^-]$ and $B = [BrMA]$, is shaded by lines of slope $= \frac{1}{2}$.
The wedge-shaped region of MSS behavior does not overlap the
accessible region of parameter space.

From Fig. 2 we conclude that for reasonable values of the chemical parameters, $A = [BrO_3^-]$ and $B = [BrMA]$, DE(4) admits only one steady state solution in the positive quadrant. Strictly speaking, we have derived this result only in the limit $\epsilon \to 0$, i.e. C = total cerium concentration $\to \infty$. However, numerical calculations (Witten, private communication) of the roots of Eq. (6) show that Fig. 1 is accurate for $\epsilon \leq 1$. For $\epsilon > 1$ the tip of the wedge moves up and to the left. For $\epsilon > 5$, Eq. (6) has three real positive roots for realistic values of A and B . However, $\epsilon > 5$ corresponds to $C < 10^{-6}M$, which is so much smaller than the total cerium concentrations for which the model was designed that it is not reasonable to trust such a prediction of MSS behavior.

If we were to consider slightly different expressions for α and δ in terms of A and B , then we could find conditions for MSS behavior. For instance, with the same $\delta = B/2A$ but a different $\alpha = (1M^{-\frac{1}{2}})\sqrt{B}(A + 0.05M)$ the accessible region of parameter space in Fig. 2 moves down and to the right, predicting MSS for $B \approx 1M$ and $A \approx 0.05M$. Othmer (1975) has studied thoroughly the qualitative properties of solutions of Eq. (4) for all values of α and δ , using methods of bifurcation theory. In the next section we shall study by phase plane techniques the properties of periodic solutions of Eq. (4) when ϵ is small and (α, δ) lies in the accessible region in Fig. 2, i.e. when there exists only one steady state in the positive quadrant. Those readers interested in more general (and more interesting!) properties of solutions of Eq. (4) are referred to Othmer's paper.

Relaxation oscillations

As in Chapter III we study periodic solutions of Eq. (4) by plotting

the nullclines, $P = 0$ and $Q = 0$. First, consider the function $x(y)$

defined implicitly by $Q(x,y;\alpha,\epsilon) = 0$, at fixed α and ϵ . See Eq. (5b).

For $y \gg \epsilon$ the relation is reciprocal parabolic

$$x = \frac{1}{1 + \alpha + (y - \alpha)^2} , \quad y \gg \epsilon .$$

This function has a local maximum at $y = \alpha$, $x = 1/(1 + \alpha)$. On the

other hand, at fixed ϵ , as $y \to 0$, $x(y) \to +\infty$. Thus there must be a

local minimum for y small. To find it we investigate the zeroes of

$$\left. \frac{dx}{dy} \right|_{Q=0} = - \frac{\epsilon[1 + \alpha + (y - \alpha)^2] + 2y(y + \epsilon)(y - \alpha)}{y^2[1 + \alpha + (y - \alpha)]^2} .$$

Besides the zero at $y = \alpha + \mathcal{O}(\epsilon)$, there is another at

$$y = \sqrt{\epsilon(1 + \alpha + \alpha^2)/2\alpha} + \mathcal{O}(\epsilon) .$$

At this value of y , $x(y) \approx 1/(1 + \alpha + \alpha^2)$, which is a local minimum.

These facts about the nullcline $Q = 0$ are illustrated in Fig. 3.

At fixed δ , the other nullcline $P = 0$, Eq. (5a), is simply an

hyperbola, which passes through the origin and is asymptotic to $x = 1$

for large y . As δ increases the hyperbola flattens. See Fig. 3.

Though it is possible for $P = 0$ and $Q = 0$ to intersect three times

(for values of α and δ within the wedge in Fig. 1), we shall assume

that there is only one intersection, i.e. only one steady state. In this

case we distinguish three possibilities: (i) for $0 < \delta < \delta_0$ the

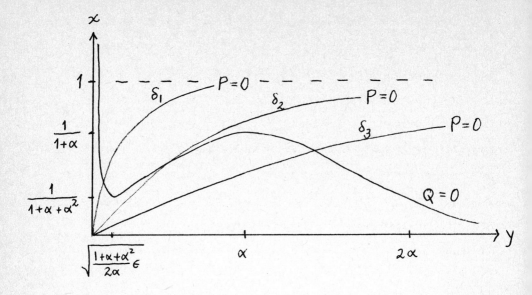

Fig. 3. The two nullclines, $P(x,y;\delta) = 0$ and $Q(x,y;\alpha,\epsilon) = 0$, at fixed α,ϵ and for three values of δ: $0 < \delta_1 < \delta_0$, $\delta_0 < \delta_2 < \delta^0$, $\delta^0 < \delta_3 < \infty$, where δ_0 and δ^0 are given by Eq. (12).

intersection lies on the steeply falling section of $Q = 0$, i.e. $0 < y \ll 1$, $(dx/dy)_{Q=0} < 0$, (ii) for $\delta_0 < \delta < \delta^0$ the intersection lies on the rising section of $Q = 0$, (iii) for $\delta^0 < \delta < \infty$ the intersection lies on the gently falling section of $Q = 0$, i.e. $y > \alpha$. By insisting that the hyperbola $x = y/(\delta + y)$ pass through either the local minimum or the local maximum of $Q = 0$ we find that

$$\delta_0 = \alpha(\alpha + 1)\sqrt{\epsilon(1 + \alpha + \alpha^2)/2\alpha},$$

(12)

$$\delta^0 = \alpha^2.$$

In the limit $\epsilon \to 0$, $y(\tau)$ changes much more rapidly than $x(\tau)$.
Except near $Q = 0$ the vector field $(\dot{x},\dot{y})$ is everywhere nearly
horizontal. The two falling sections of the one-dimensional manifold
$Q = 0$ are stable, but the middle section is unstable. (We referred to
this fact earlier.) For $0 < \delta < \delta_0$ and $\delta^0 < \delta < \infty$, we find that the
steady state is globally asymptotically stable (as $\epsilon \to 0$). However,
under these conditions the system is excitable in the sense described in
Chapter IV (pp. 76f). For $\delta_0 < \delta < \delta^0$ we find an orbitally asymptotically stable periodic solution illustrated in Fig. 4.

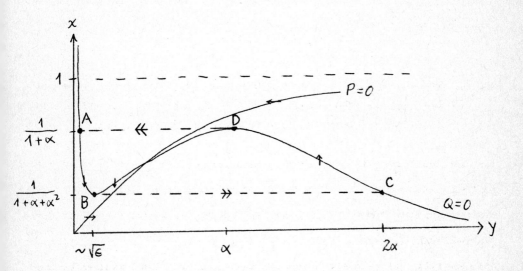

Fig. 4. Discontinuous periodic solution (ABCD) of Eq. (4) in the limit
$\epsilon \to 0$ for $\delta_0 < \delta < \delta^0$.

Since this is not a rigorous analysis, the predictions may not be
accurate for $\delta \approx \delta_0$ and $\delta \approx \delta^0$. Othmer (1975) plots some numerically

calculated limit cycles for $\alpha = 4$, $\delta = 2.5$, 3.15, 3.35, 5.0. Fig. 4 is quantitatively accurate only in the last case. However, the simple description derived here gives some insight into the behavior of solutions of Eq. (4) when the steady state solution is unique. For instance, Table 1 compares the maximum and minimum Ce^{+4} concentrations measured by ZZKK ,

Table 1. Observed, calculated, and predicted Ce^{+4} concentrations[*].

		B = [BrMA] in moles/liter				
		0.005	0.01	0.032	0.1	0.32
A = [BrO$_3^-$] in moles/liter	0.01	55, 35 54, 38 51, 34	43, 23 45, 25 42, 23	31, 9 32, 11 29, 11	20, 3 22, 3.7 19, 4.1	
	0.05		78, 72 74, 71 73, 66	65, 47 61, 50 60, 47	52, 26 47, 28 46, 28	34, 10 34, 12 32, 13
	0.1			75, 68 71, 65 69, 61	64, 44 57, 43 56, 41	40, 17 43, 22 41, 23

[*]In each box of data is reported:

(i)   in the first line, the percentage of cerium in the +4 oxidation state at the maximum and minimum points of observed oscillations (Zhabotinskii, et. al., 1971). Experimental conditions: $40°C$, $6MH_2SO_4$, total cerium = 10^{-3} moles/liter.

(ii) in the second line, the same percentages calculated by numerical integration of Eq. (4), using Eq. (10) to calculate parameters (Zhabotinskii, et. al, 1971).

(iii) in the third line, the same percentages as predicted by Fig. 4, i.e. $100/(1 + \alpha)$ at the maximum and $100/(1 + \alpha + \alpha^2)$ at the minimum.

calculated by numerical integration of Eq. (4) and predicted by the discontinuous oscillation in Fig. 4[*].

Fig. 5 compares the region of oscillations in parameter space predicted by Eq. (12), calculated by Othmer (1975) and measured by ZZKK . Here the agreement is less satisfactory.

---

[*]ZZKK also compare the observed period with their numerical calculations. Unfortunately, we cannot get a simple expression for the period from Fig. 4 because integrals of the form

$$\tau_{CD} = \int_C^D \frac{dx}{(dx/d\tau)_{Q=0}} = \int_0^{-\alpha} \frac{2u\,du}{(u^2 + \alpha + 1)(u^3 + \alpha u^2 + \alpha u + \alpha^2 - \delta)} \, ,$$

where $u = -\alpha + \sqrt{x^{-1} - (1 + \alpha)}$, cannot be evaluated in closed form.

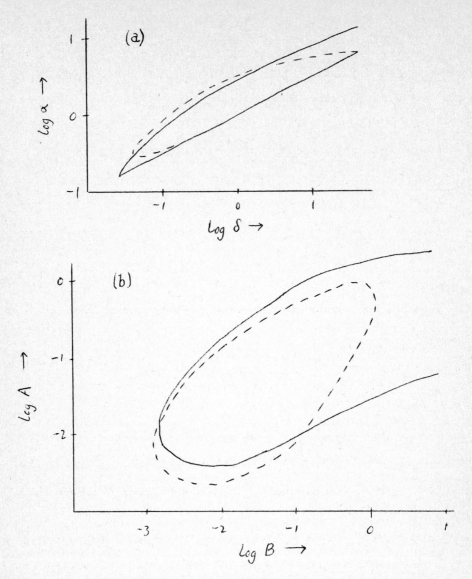

Fig. 5. (a) Stability of the unique steady state as predicted by Eq. (12)
when $\epsilon \to 0$ (———) and as computed by Othmer (1975) for
$\epsilon = 5 \times 10^{-3}$ (---) . (b) Values of $A = [BrO_3^-]$, $B = [BrMA]$ for
which ZZKK (1971) observed oscillations (---) and for which
Eq. (4) predicts oscillations for $\epsilon \to 0$ (———) . The solid curves
in (a) and (b) are identical and related by Eq. (10).

REFERENCES

1) Andronov, A. A., Vitt, A. A. and Khaikin, S. E. (1966). Theory of Oscillators (Pergamon, London).

2) Aronson, D. G. and Weinberger, H. F. (1975). "Nonlinear diffusion in population genetics, combustion, and nerve pulse propagation." In J. A. Goldstein, Ed. Partial Differential Equations and Related Topics, Lecture Notes in Mathematics, Vol. 446 (Springer, Berlin), pp. 5-49.

3) Belousov, B. P. (1958). Sb. Ref. Radiat. Med. (Collections of Abstracts on Radiation Medicine, Medgiz, Moscow), p. 145.

4) Bornmann, L., Busse, H. and Hess, B. (1973a). "Oscillatory oxidation of malonic acid by bromate. 1. Thin-layer chromatography." Z. Naturforsch. 28b, 93-97.

5) _____ (1973b). "Oscillatory oxidation of malonic acid by bromate. 3. CO_2 and Br^- titration." Z. Naturforsch. 28c, 514-516.

6) Bourret, J. A., Lincoln, R. G. and Carpenter, B. H. (1969). "Fungal endogenous rhythms expressed by spiral figures." Science 166, 763-765.

7) Boyce, W. E. and DiPrima, R. C. (1969). Elementary Differential Equations and Boundary Value Problems, 2nd ed. (Wiley, New York).

8) Bray, W. C. and Liebhafsky, H. A. (1935). "The kinetic salt effect on the fourth order reaction $BrO_3^- + Br^- + 2H^+ \rightarrow$." J. Am. Chem. Soc. 57, 51-56.

9) Burnside, W. S. and Panton, A. W. (1928). The Theory of Equations, Vol. I, 9th ed. (Longmans, Green and Co., London).

10) Busse, H. (1969). "A spatial periodic homogeneous chemical reaction." J. Phys. Chem. 73, 750.

11) Chance, B., Ghosh, A. K., Pye, E. K. and Hess, B. (1973). Biological and Biochemical Oscillators (Academic, New York).

12) Coddington, E. A. and Levinson, N. (1955). Theory of Ordinary Differential Equations (McGraw-Hill, New York).

13) Daniels, F. and Alberty, R. A. (1966). Physical Chemistry, 3rd. ed. (J. Wiley, New York).

14) Degn, H. (1967). "Effect of bromine derivatives of malonic acid on the oscillating reaction of malonic acid, cerium ions and bromate." Nature 213, 589-590.

15) DeSimone, J. A., Beil, D. L. and Scriven, L. E. (1973). "Ferroin-collodion membranes: dynamic concentration patterns in planar membranes." Science 180, 946-948.

16) Edelson, D., Field, R. J. and Noyes, R. M. (1975). "Mechanistic details of the Belousov-Zhabotinskii oscillations." Int. J. Chem. Kin. 7, 417-432.

17) Edelstein, B. B. (1970). "Biochemical model with multiple steady states and hysteresis." J. Theor. Biol. 29, 57-62.

18) Field, R. J. (1972). "A reaction periodic in time and space. A lecture demonstration." J. Chem. Educ. 49, 308-311.

19) _____ (1975). "Limit cycle oscillations in the reversible oregonator." J.Chem.Phys. 63, 2284-2296.

20) Field, R. J., Körös, E. and Noyes, R. M. (1972). "Oscillations in chemical systems. II. Thorough analysis of temporal oscillations in the bromate-cerium-malonic acid system." J. Am. Chem. Soc. 94, 8649-8664.

21) Field, R. J. and Noyes, R. M. (1972). "Explanation of spatial band propagation in the Belousov reaction." Nature 237, 390-392.

22) ———————————————— (1974a). "Oscillations in chemical systems. IV. Limit cycle behavior in a model of a real chemical reaction." J. Chem. Phys. $\underline{60}$, 1877-1884.

23) ———————————————— (1974b). "Oscillations in chemical systems. V. Quantitative explanation of band migration in the Belousov-Zhabotinskii reaction." J. Am. Chem. Soc. $\underline{96}$, 2001-2006.

24) Fisher, R. A. (1937). "The wave of advance of advantageous genes." Ann. Eugenics $\underline{7}$, 355-369.

25) FitzHugh, R. (1961). "Impulses and physiological states in theoretical models of nerve membrane." Biophys. J. $\underline{1}$, 445-466.

26) Gantmacher, F. R. (1959). The Theory of Matrices, Vol. II. (Chelsea Co., New York).

27) Greenberg, J. M. (1975). "Periodic solutions to reaction/diffusion equations." SIAM J. Appl. Math., in press.

28) Gul'ko, F. B. and Petrov, A. A. (1972). "Mechanism of the formation of closed pathways of conduction in excitable media." Biophysics $\underline{17}$, 271-282.

29) Hastings, S. P. (1975). "Some mathematical problems from neural biology." Am. Math. Monthly $\underline{82}$, 881-895.

30) Hastings, S. P. and Murray, J. D. (1975). "The existence of oscillatory solutions in the Field-Noyes model for the Belousov-Zhabotinskii reaction." SIAM J. Appl. Math. $\underline{28}$, 678-688.

31) Hirsch, M. W. and Smale, S. (1974). Differential Equations, Dynamical Systems and Linear Algebra (Academic, New York).

32) Hopf, E. (1942). "Abzweigung einer periodischen Lösung von einer stationären Lösung eines Differential systems." Ber. Math.-Phys. Klasse Sächs. Akad. Wiss. Leipz. $\underline{94}$, 3-22.

33) Hsü, I.-D. and Kazarinoff, N. D. (1975). "An applicable Hopf bifurcation formula and instability of small periodic solutions of the Field-Noyes model." J. Math. Anal. Appl., in press.

34) Kasperek, G. J. and Bruice, T. C. (1971). "Observations on an oscillatory reaction. The reaction of potassium bromate, ceric sulphate, and a dicarboxylic acid." Inorg. Chem. $\underline{10}$, 382-386.

35) Kolmogorov, A., Petrovsky, I. and Piscounoff, N. (1937). "Étude de l'équation de la diffusion avec croissance de la quantité de matière et son application à un problème biologique." Bull. Univ. Moskou (Ser. internat.) $\underline{A1}$, 1-25.

36) Kopell, N. and Howard, L. N. (1973a). "Horizontal bands in the Belousov reaction." Science $\underline{180}$, 1171-1173.

37) _____ (1973b). "Plane wave solutions to reaction-diffusion equations." Studies Appl. Math. $\underline{52}$, 291-328.

38) _____ (1974). "Pattern formation in the Belousov reaction." In S. Levin, Ed. Lectures on Mathematics in the Life Sciences, vol. 7 (Am. Math. Soc., Providence, R. I.), pp. 201-216.

39) _____ (1975). "Waves in an oscillatory chemical medium II: Diffusion waves." SIAM Rev., to appear.

40) Krinsky, V. I. (1973). "Excitation wave propagation during heart fibrillation." In B. Chance, et al., Eds., op. cit., pp. 329-341.

41) Lavenda, B., Nicolis, G. and Herschkowitz-Kaufman, M. (1971). "Chemical instabilities and relaxation oscillations." J. Theor. Biol. $\underline{32}$, 283-292.

42) Marsden, J. and McCracken, M., Eds. (1975). The Hopf Bifurcation, Lecture Notes in Mathematics, to appear.

43) McKean, Jr., H. P. (1970). "Nagumo's equation." Adv. in Math. $\underline{4}$, 209-223

44) Minorsky, N. (1964). "Nonlinear problems in physics and engineering." In H. Margenau, G. M. Murphy, Ed. The Mathematics of Physics and Chemistry, Vol. II. (D. van Nostrand, Princeton, N. J.), p. 321-390.

45) Murray, J. D. (1974a). "On a model for temporal oscillations in the Belousov-Zhabotinskii reaction." J. Chem. Phys. 61, 3610-3613.

46) _____ (1974b). "On a model for concentration waves in the Belousov-Zhabotinskii reaction." J. Theor. Biol., in press.

47) Nagumo, J., Arimoto, S. and Yoshizawa, S. (1962). "An active pulse transmission line simulating nerve axon." Proc. I.R.E. 50, 2061-2070.

48) Nicolis, G. (1971). "Stability and dissipative structures in open systems far from equilibrium." Adv. Chem. Phys. 19, 209-324.

49) Nicolis, G. and Portnow, J. (1973). "Chemical oscillations." Chem. Rev. 73, 365-384.

50) Noyes, R. M. (1976). "Oscillations in chemical systems. XII. Applicability to closed systems of models with two and three variables." J. Chem. Phys. 64, 1266.

51) Noyes, R. M. and Jwo, J.-J. (1975). "Oscillations in chemical systems. X. Implications of cerium oxidation mechanisms for the Belousov-Zhabotinskii reaction." J. Am. Chem. Soc. 97, 5431-5433.

52) Ortoleva, P. and Ross, J. (1974). "On a variety of wave phenomena in chemical reactions." J. Chem. Phys. 60, 5090-5107.

53) Othmer, H. G. (1975). "On the temporal characteristics of a model for the Zhabotinskii-Belousov reaction." Math. Biosci. 24, 205-238.

54) Prigogine, I. and Lefever, R. (1968). "Symmetry breaking instabilities in dissipative systems. II." J. Chem. Phys. 48, 1695-1700.

55) Rinzel, J. and Keller, J. B. (1973). "Travelling wave solutions of a nerve conduction equation." Biophys. J. 13, 1313-1337.

56) Robertson, A. and Cohen, M. H. (1972). "Control of developing fields." Ann. Rev. Biophys. Bioeng. 1, 409-464.

57) Schmitz, R. A. (1974). "Multiplicity, stability and sensitivity of states in chemically reacting systems--a review." In Third International Symposium on Chemical Reaction Engineering, Advances in Chemistry Series (Am.' Chem. Soc., Washington, D. C.), to appear.

58) Smith, R. H. (1972). "Rate constants for the oxidation of formate and oxalate by aqueous bromine." Aust. J. Chem. 25, 2503-2506.

59) Stanshine, J. A. (1975). "Asymptotic solutions of the Field-Noyes model for the Belousov reaction." Thesis, Mass. Inst. Technology.

60) Sullivan, J. H. (1967). "Mechanism of the 'bimolecular' hydrogen-iodine reaction." J. Chem. Phys. 46, 73-78.

61) Tatterson, D. F. and Hudson, J. L. (1973). "An experimental study of chemical wave propagation." Chem. Eng. Commun. 1, 3-11.

62) Thoenes, D. (1973). "'Spatial oscillations' in the Zhabotinskii reaction." Nature (Phys. Sci.) 243, 18-20.

63) Tilden, J. (1974). "On the velocity of spatial wave propagation in the Belousov reaction." J. Chem. Phys. 60, 3349-3350.

64) Troy, W. C. and Field, R. J. (1975). "The amplification before decay of perturbations around stable states in a model of the Zhabotinskii reaction." SIAM J. Appl. Math., to appear.

65) Tyson, J. J. (1973). "Some further studies of nonlinear oscillations in chemical systems." J. Chem. Phys., 58, 3919-3930.

66) _____ (1975). "Analytic representation of oscillations, excitability and traveling waves in a realistic model of the Belousov-Zhabotinskii reaction." J. Chem. Phys., in press.

67) _____ (1976). "Analytic representation of multiple steady states, oscillations and excitability in a model of the Belousov-Zhabotinskii reaction." In preparation.

68) Vavilin, V. A. and Zhabotinskii, A. M. (1969). "Induced oxidation of tribromoacetic and dibromomalonic acids." Kinetics and Catalysis 10, 538-540.

69) West, R. W. (1924). "The action between bromine and malonic acid in aqueous solution." J. Chem. Soc. 125, 1277-1282.

70) Winfree, A. T. (1972). "Spiral waves of chemical activity." Science 175, 634-636.

71) _____ (1973). "Scroll-shaped waves of chemical activity in three dimensions." Science 181, 937-939.

72) _____ (1974a). "Rotating solutions to reaction/diffusion equations in simply-connected media." In D. S. Cohen, Ed., Mathematical Aspects of Chemical and Biochemical Problems and Quantum Chemistry, SIAM-AMS Proceedings, Vol. 8. (Amer. Math. Soc., Providence, R. I.).

73) _____ (1974b). "Rotating chemical reaction." Sci. Amer. 230, 82-95.

74) _____ (1974c). "Two kinds of wave in an oscillating chemical solution." In the Faraday Symposium of the Chemical Society, No. 9, pp. 38-46.

75) Zaikin, A. N. and Zhabotinskii, A. M. (1970). "Concentration wave propagation in two-dimensional liquid-phase self-oscillating system." Nature 225, 535-537.

76) Zhabotinskii, A. M. (1964). "Periodic course of oxidation of malonic acid in solution (investigation of the kinetics of the reaction of Belousov)." Biophysics 9, 329-335.

77) Zhabotinskii, A. M., Zaikin, A. N., Korzukhin, M. D. and Kreitser, G. P. (1971). "Mathematical model of a self-oscillating chemical reaction (oxidation of bromomalonic acid with bromate, catalyzed by cerium ions)." Kinetics and Catalysis 12, 516-521.

Editors: K. Krickeberg;
S. Levin; R. C. Lewontin;
J. Neyman; M. Schreiber

Biomathematics

Springer-Verlag
Berlin
Heidelberg
New York